THE Aviator's Catalog

N8251P

THE Aviator's Catalog

A SOURCE BOOK OF AERONAUTICA

Timothy R. V. Foster

VNR VAN NOSTRAND REINHOLD COMPANY
New York Cincinnati Toronto London Melbourne

Also by Timothy R. V. Foster:

The Aircraft Owner's Handbook: Everything You Need to Know about Buying, Operating and Selling an Aircraft (Van Nostrand Reinhold Company, 1978)

Library of Congress Catalog
Card Number 80-10222
ISBN 0-442-21201-1 (cloth)
ISBN 0-442-22465-6 (paper)

Printed in the United States of America.

Designed by Jean Callan King/Visuality.

Published by Van Nostrand Reinhold Company Inc.
135 West 50th Street, New York, NY 10020

Fleet Publishers
1410 Birchmount Road
Scarborough, Ontario M1P 2E7, Canada

Van Nostrand Reinhold
48 Latrobe Street
Melbourne, Victoria 3000, Australia

Van Nostrand Reinhold Company Limited
Molly Millars Lane
Wokingham, Berkshire RG11 2PY, England

16 15 14 13 12 11 10 9 8 7 6 5 4 3

Library of Congress Cataloging in Publication Data

Foster, Timothy R V
The aviator's catalog.
Includes index.
1. Airplanes—Equipment and supplies—Catalogs.
2. Private flying—United States—Miscellanea.
I. Title
TL688.F67 629.13'0973 80-10222
ISBN 0-442-21201-1
ISBN 0-442-22465-6 pbk.

CONTENTS

For Lauren Patricia

Acknowledgments

Before I run through the usual long list of names of the people who helped me put this together, I want to acknowledge my typewriter, which is, without a doubt, the greatest system available to a compiler of complex data. It's a Lanier *No Problem* computer word-processor. Even though I am one of the world's fastest and most inaccurate typists, this device has enabled me to create this book virtually single-handed with nary a typographical "erorr." Not only that, it's fun to use.

These people were among the hundreds who helped me put this book together and I would like to acknowledge them:

Joanne Alford, John Alter, Marty Balk, Dick Baughman, Bob Beach, Norm Bender, Karen Borger, Ed Braden, Morgan Brown, Brian Calvert, John Castelli, Craig Christie, Marion Cole, Keith Connes, Karen Coyle, Bob Dalin, Jack Elliott, John Ferrara, Mary Foster, Rudy Frasca, Ed Gray, Jim Gregory, Kitty Howser, Holly Hudlow, Dean Humphrey, Jay Lavenson, Alton Lewis, Howard Levy, Norman Lornie, Joanne Lubitz, Martha Lunken, Basil Maile, Sherry Marshall, Phil McCoy, Bernie McGowan, John Meyer, Curvin O'Rielly, Joe Robinson, Cole Palen, Bob Rawe, Ken Ross, Laurie Schwartz, Norbert Slepyan, Marvin Stern, Ed Stimpson, Don Stretch, Molt Taylor, Richard Uppstrom, Charlie Vogel, Dick Weeghman, Leslie M. Wenger, Ken Wickham, John Zimmerman, Herman Zollar.

PREFACE

Aviators are among the world's greatest gadget lovers. They love to indulge, nay immerse themselves, in things aeronautical. Fortunately, there is a dynamic industry at work serving the needs of pilots and aviation buffs. If you're one of these (and probably both), this book is designed to help you live with your needs. It may even help you find true fulfillment. It shows you what the latest paraphernalia and gadgets are and where to get them. It gives you much other useful and relevant information.

Think of *The Aviator's Catalog* as an encyclopedia of useful aeronautical material. It is organized into six broad and completely arbitrary general areas: *Flying, Aircraft, Aircraft Operation, Aviator's Equipment and Services, Aerobuff Stuff,* and *Etc.*

To help you find your way, there is a comprehensive table of contents and a very complete index. For example, if you are looking for a leather flying helmet, you could look in the index under "helmets, flying," which will direct you to the appropriate section. Or you can leaf through the pages until you find it. If you do the latter, you're bound to find something else of interest, such as where to find a french-fried-potato stand made from an old Lockheed Lodestar, or which aircraft has the highest seat-miles-per-gallon.

By the way, this book is *not* a mail-order catalog. It is a *source book.* You can't order anything from it, but you will find out where to go to get what you want.

A Word About Prices

Obviously, prices change, and I'm afraid I have found there is little *I* can do about inflation—God knows I've tried! So my solution has been to show *approximate prices,* which were roughly accurate at the time of writing. Since the suppliers' names, addresses, and phone numbers are given in almost all listings, I urge you to check with them first before ordering anything. Don't order from the publisher of this book! Go to the source shown or a substitute you know about.

P A R T O N E

Flying

PILOT CERTIFICATES

LEARNING TO FLY

INSTRUMENT RATING

FLIGHT SIMULATORS

FLYING CLUBS

AEROBATICS

WATER FLYING

GLIDING

HANG GLIDING

BALLOONING

Piper's Turbo Seminole—the state-of-the-art in light, light twins. (Piper photo)

Pilot Certificates

To fly an airplane, you must have a pilot certificate or license. All licenses throughout the free world are issued in accordance with requirements of the **International Civil Aviation Organization, Box 400, Montreal, PQ, H3A 2R2, Canada, telephone (514) 285-8219,** but some minor exceptions may exist between nations.

The FAA rules governing pilot licensing are in FAR (Federal Aviation Regulations) Part 61. Basic requirements are that you must have a valid FAA pilot certificate and a currently valid medical certificate to operate as a pilot in command of an aircraft within the United States. However, you may fly a foreign-registered aircraft in the United States if you hold a license issued by the country in which the aircraft is registered. You must be able to read, speak, and understand English to obtain an FAA certificate.

Pilot certificates are as follows:

- Student pilot
- Private pilot
- Commercial pilot
- Airline transport pilot

In addition there are flight instructor certificates and a variety of ratings:

Aircraft *category* ratings:

- Airplane
- Rotorcraft
- Glider
- Lighter-than-air

Airplane *class* ratings:

- Single-engine land
- Multi-engine land
- Single-engine sea
- Multi-engine sea

Rotorcraft class ratings:

- Helicopter
- Gyroplane

Lighter-than-air class ratings:

- Airship
- Free balloon

Aircraft-type ratings are covered in FAA Advisory Circular 61-1 "Aircraft Type Ratings," obtainable free from **The Department of Transportation, Publications Section, M-443.1, Washington, DC 20590.** Ratings available include:

- Large aircraft, other than lighter-than-air
- Small turbojet-powered airplanes
- Small helicopters for operations requiring an airline-transport-pilot certificate
- Other aircraft-type ratings, as required by type certificate

Instrument ratings (on private and commercial certificates only—the airline transport certificate includes an instrument rating):

- Instrument—airplane
- Instrument—helicopter

The following ratings are placed on flight-instructor certificates where applicable:

Aircraft category ratings:

- Airplane
- Rotorcraft
- Glider

Airplane class ratings:

- Single-engine
- Multi-engine

Rotorcraft class ratings:

- Helicopter
- Gyroplane

Instrument ratings:

- Instrument—airplane
- Instrument—helicopter

Medical certificates

Requirements for medical certificates are given in the following FARs: 61.3a, 61.23, 61.29b, 61.53, 61.75f, 61.83c, 61.103c, 61.123c, and 61.151e.

All pilots, except free-balloon pilots piloting balloons and glider pilots piloting gliders, must have a currently valid medical certificate in their possession when they are exercising the privileges of their certificate (see FAR 61.3a). There are three classes of medical certificate (see FAR 61.23):

- First class, required for an airline transport pilot certificate. This is valid for six months.
- Second class, required for a commercial pilot certificate. This is valid for one year.
- Third class, required for a private or student pilot certificate. This is valid for two years.

If you hold a first-class medical certificate, you can use it for second-class privileges for up to one year, and for third-class privileges for up to two years. If you hold a second-class certificate, you can use it for third-class privileges for up to two years.

If you lose your medical certificate, you can obtain a replacement by writing: **FAA Aeromedical Certification Branch, Box 25082, Oklahoma City, OK 73125,** and enclosing a check or money order for $2. A telegraphic certificate is available for rush jobs (see FAR 61.29b,c).

No person may act as pilot in command, or in any other capacity as a required flight crewmember, while suffering a known medical deficiency, or increase of a known deficiency, that would preclude them from meeting the requirements of their current medical certificate (see FAR 61.53).

Experience requirements for pilot certificates

The following table outlines certain minimum experience requirements. However, there are many rinky-dinks that modify these requirements. For example, FAA-approved courses have lower requirements than unapproved courses; 50 hours of the total flight time for the commercial may be done in ground trainers; some of the flight time for an ATP may be flight-engineer time; pilots not meeting certain night, cross-country, or instrument-time requirements may still obtain certificates, but with a limiting endorsement, and so on. For full details, check the appropriate FAR subpart indicated in the table, using the latest FARs published by the FAA.

■ **TABLE OF MINIMUM EXPERIENCE AND OTHER REQUIREMENTS FOR FAA PILOT CERTIFICATES—AIRPLANE CATEGORY**

	Student	Private	Commercial	ATP
FAR Part 61 Subpart ref.	C	D	E	F
Minimum age	16	17	18	23
Medical class	3rd	3rd	2nd	1st
Flight-time requirements in hours				
Total	None	40	250	1500
Instruction	None	20	50	50
Solo/command	None	20	100	250
Instrument instruction	None	None	20	20
Total instrument	None	None	40	75
Total XC	None	13	53	500
Command XC	None	10	50	50
Total night	None	3	5	100

Written exams

Most certificates and some ratings require the applicant to pass a written test. These tests are all multiple choice, and the FAA has recently published all the questions on their exams—thousands of them. You only get maybe 100 or so questions in the exam you take; the questions are selected from this vast array. Before you take any written exam, you must show that you have satisfactorily completed the ground-instruction or home-study course required for the exam you want to take, and you must show identification and proof of age. You can in fact be up to two years *younger* than the minimum age for the license sought, since a passed exam is valid for two years for completion of the other licensing requirements. This is all spelled out in FAR 61.35.

Here are the FAA pilot certificates and ratings that require a written test:

- Private Pilot Certificate
- Commercial Pilot Certificate
- Airline Transport Pilot Certificate
- Flight Instructor Certificate
- Instrument Rating
- Certificate based on a foreign license
- Certificate based on military pilot qualifications

A lot of study goes into getting your license. This Jeppesen kit is typical of the material used. (Jeppesen-Sanderson photo)

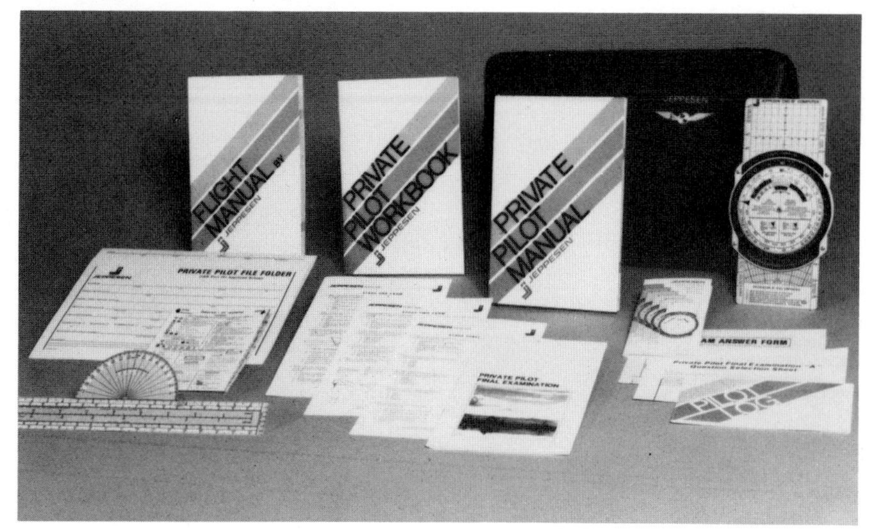

A given test covers the aircraft category for which the certificate is sought, so if you're seeking a helicopter endorsement, your written test must cover helicopters. If you are to become an airship pilot, your written covers lighter-than-air dirigibles, and so on.

If you take your test at an FAA General Aviation District Office (GADO), the most important thing to remember is that you must be there with enough time available in the day to give you the full time allowed for the test, and still get out before they close, usually at 4:30 PM. Thus, if it's a four-hour exam, you have to be there before 12:30 PM.

Flight tests

A flight test, administered by an FAA Inspector or FAA-authorized Examiner, must be passed for the issuance of the following certificates and ratings:

- Private Pilot Certificate
- Commercial Pilot Certificate
- Airline Transport Pilot Certificate
- Flight Instructor Certificate
- Instrument Rating
- Multi-engine Rating

To be eligible to take the flight test, you must have passed the applicable written exam within the preceding 24 months, hold an appropriate current medical certificate, be the appropriate age, and have a letter from a flight instructor saying you've prepared for the test and are ready to take it (the letter and age requirements don't exist for an ATP flight test). See FAR 61.39 for full details.

FAA study guides

The FAA publishes written test and flight test study guides in the form of Advisory Circulars:

■ WRITTEN TEST GUIDES

Title	Circular Number
FAR Written Test Guide for Private, Commercial and Military Pilots	AC 61-34B
Private Pilot, Airplane	AC 61-32A
Instrument Pilot	AC 61-8C
Commercial Pilot, Airplane	AC 61-71
Private and Commercial Pilot (Rotorcraft-Helicopter)	AC 61-73
Flight Instructor, Airplane	AC 61-72
Flight Instructor, Instrument	AC 61-70
Flight Instructor, Rotorcraft-Helicopter	AC 61-74
Flight Instructor, Glider	AC 61-75
Airline Transport Pilot, Airplane	AC 61-18D
Airline Transport Pilot, Helicopter	AC 61-42A

The Jeppesen Commercial/Instrument study guides are popular with advanced students. (Jeppesen-Sanderson photo)

■ FLIGHT TEST GUIDES

Title	Circular Number
Private Pilot, Airplane	AC 61-54A
Instrument Pilot, Airplane	AC 61-56A
Commercial Pilot, Airplane	AC 61-55A
Flight Instructor, Airplane	AC 61-58
Airline Transport Pilot, Airplane	AC 61-77
Type Rating, Airplane	AC 61-57A
Private and Commercial, Helicopter	AC 61-59
Instrument Pilot, Helicopter	AC 61-64
Private and Commercial, Gyroplane	AC 61-60
Private and Commercial, Glider	AC 61-61

These FAA Advisory Circulars may be bought at most fixed base operators (FBO's) and government book stores.

Pilot certificate privileges

The privileges for each grade of pilot certificate are given in the appropriate subpart of FAR 61. Basically, a student pilot may not carry passengers and may fly only under the direction of an authorized flight instructor; a private pilot may not carry passengers or property for hire, although expenses may be shared; a commercial pilot may act as pilot in command of an aircraft carrying persons or property for hire; an airline-transport pilot has the privileges of a commercial pilot with an instrument rating; an instrument rating entitles a pilot to fly according to the instrument flight rules (IFR).

Learning to Fly

The process of flight training has become quite sophisticated in the past few years, chiefly due to the establishment of aircraft manufacturers' franchised flight schools. The big three manufacturers all promote their own franchised flight-training centers, which use audio-visual aids for self-paced learning and a highly organized curriculum to make the best use of the time available. Each manufacturer offers a free booklet that provides the basic information about its courses. The Cessna Pilot Centers have "The Answer Book," available from **Cessna Aircraft Company, Wichita, KS 67201, telephone (316) 685-9111.** This not only gives a lot of facts, but also includes a certificate that will get you a trial lesson for a low fee. Piper Flite Centers offer a booklet, "This is the Sum Total of Everything Ever Learned About Learning to Fly," available from **Piper Aircraft Corporation, Lock Haven, PA 17745, telephone (717) 748-6711.**

The Gulfstream American Trainer is out of production but can still be found at some flight schools. (Gulfstream American photo)

Beech Aero Clubs offer the "Beech Aero Center Adventure Kit," available from **Beech Aircraft Corporation, Wichita, KS 67201, telephone (316) 681-7111.**

Gulfstream American also has a flight school operation called the Gulfstream American Flying Center. However, since GA's parent, American Jet Industries, Inc., has sold their single-engine line, the future of these centers is in the air, if you'll pardon the pun. **Gulfstream American Corporation, Box 2206, Savannah, GA 31402, telephone (912) 964-3000.**

Other schools

There are many other flight schools not affiliated with any particular manufacturer. All schools must be approved by the FAA to offer flight training under FAR Part 141, which gives lower minimum requirements for the issuance of a pilot certificate. However, schools do not *have* to be FAA approved, in which case the minimum experience requirements are higher. To find the location of your nearest flight school, look in the *Yellow Pages* under "Aircraft Schools," or go to your local airport and sniff around.

Top: The Piper Tomahawk trainer is the mainstay of the Piper Flite Center fleet. (Piper photo) *Middle*: The world's best-selling trainer—the Cessna 152. (Cessna photo) *Bottom*: Beechcraft's new Skipper joins the Beech Aero Club fleet. (Beech photo)

Costs

To get a private certificate costs about $1,200 to $1,500. Piper offers its "Blue Sky Course," which gets you to the solo stage for a low payment, as low as $300 in some areas. Schools charge separate rates for dual and solo flights—dual (with an instructor) costs about $12 to $15 an hour more than solo. Solo rates run between $20 and $30 an hour for a typical two-seater trainer, such as a Cessna 152, Piper Tomahawk, or Beech Skipper. You can pay as you go, but some schools will give you a nice discount if you pay for your lessons in advance. If you *buy* a new aircraft, many dealers will teach you to fly "free." This doesn't mean *completely* free—just that the cost of the instructor is free. You pay for the cost of the airplane, gas, maintenance, insurance, etc. Some people learn to fly on business trips, using the aircraft to get from A to B, and sneaking in some lessons before or after the trip. This can be a good way to squeeze in flying lessons if you're one of those people with no time to spare.

Time

Getting your private license requires a minimum of 35 hours of flight time on an FAA-approved course, or 40 otherwise; either way, you must have at least 20 hours of flight instruction from an authorized flight instructor. Those are *minimum* figures. Most people take considerably more time—perhaps 50 or 60 hours. This time is usually taken in lessons of about one hour's duration, so the time it takes to actually learn depends on the time you are able to commit. Theoretically, you could learn to fly and get a license in a month, if you go at it every day and have good weather. If this is your goal, your best bet is to go to one of the specialized flight schools and just fly full time. Some schools offer special "learn-to-fly-on-vacation plans," which get you up to the solo stage in a couple of weeks, and then let you continue at a slightly less hectic pace thereafter.

Most people take between six months and a year to get a license—they go solo within the first month or two and just plug away until they are ready for the flight test. Meanwhile, they have taken some fairly extensive ground instruction, carried out 3 hours of dual and 10 hours of solo cross-country flying, including the big solo cross-country that requires a flight of at least 300 miles, with landings at three points at least 100 miles away from each of the other two points. Three hours of night instruction are also required, if the license is to be valid for night flying.

A good way to log your time and keep tax records is with the *Pilot's Taxlog*. (Illustration by Pilot's Taxlog)

Insurance aspects of learning to fly

If you are taking flying lessons, check your life and accident-protection insurance coverage. Quite often you will find that it does not cover you for training flights or for flights where you are the pilot-in-command. If this is the case, arrange with the company to *make* it cover you—this might cost a small extra premium with some policies.

I won't even discuss insurance with a company that wants to charge me extra because I am a pilot. However, many group policies have a blanket anti-flying clause, and there is nothing you can do about it except arrange your own extra coverage. It's about time the insurance companies recognized that aviation is here. I have been able to arrange all my own life and accident insurance with full aviation coverage at no extra premium. My insurance agent has standing orders in this regard.

As far as hull and liability insurance on the aircraft you train in is concerned, ask the flight school what their policy is. Most schools will cover you fully in the event of an accident, with no deductible, if you are undergoing flight training, dual or solo. Once you have a license, there is usually a deductible—typically $1,000, if you prang one of their rented aircraft. There may be no liability coverage that protects your own interests. I recommend you carry renter pilot insurance to provide your own liability and hull coverage if you rent airplanes. This is available from many aviation insurance sources, such as **Avemco Insurance Company, 7315 Wisconsin Avenue, Bethesda, MD 20014, telephone** in **Maryland (301) 656-3044,** or **(800) 638-8440** otherwise east of the Mississippi River, and **(800) 433-1750** west thereof, except

Another form of insurance is the constant honing of your skills. This is an ATC 610K flight simulator. One insurance company—Avemco—even sells these simulators. (ATC photo)

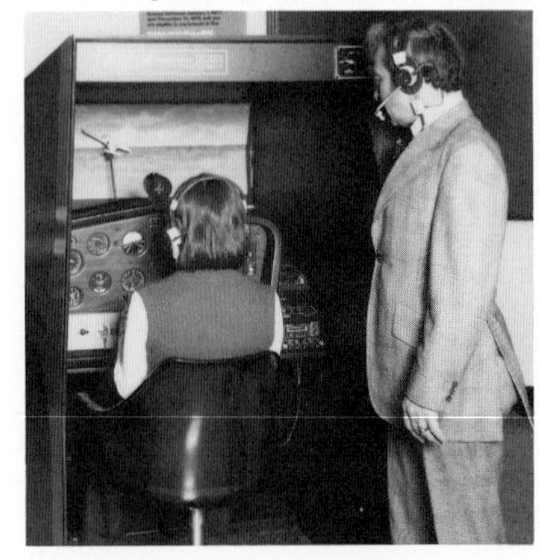

Texas, in which case call **(800) 792-1261**; or **National Aviation Underwriters, Box 10155, St. Louis, MO 63172, telephone (314) 426-2222,** or **(800) 325-8988** outside Missouri.

Instrument Rating

Without exaggeration, the most important rating available to a pilot, from a safety and utility point of view, is the instrument rating (see FAR 61.65). With this you are permitted to fly in instrument meteorological conditions (IMC) in accordance with the instrument flight rules (IFR). Otherwise, you have to fly in accordance with the visual flight rules (VFR). See the box for VFR weather minima, p. 16.

VFR minima at airports with control zones are 1,000 feet ceiling and 3 miles ground visibility. Flights may operate "Special VFR" with one mile ground visibility and "clear of clouds," with a clearance from the control tower.

Instrument rating experience requirements

To gain an instrument rating, you must hold a current private or commercial pilot certificate (the ATP *includes* an instrument rating). You must pass a written test and have a total of 200 hours of pilot time, with 100 as pilot in command and 50 of these as cross-country hours in the category of aircraft (i.e., airplane or helicopter) for which you want the rating. You must have 40 hours of simulated or actual instrument time, of which no more than 20 may be instruction given by an authorized instrument instructor in a ground trainer, and of which 15 must be instrument instruction given by an authorized instrument flight instructor,

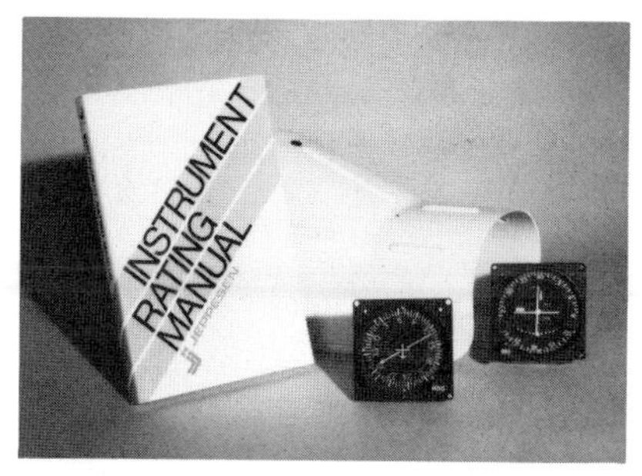

A study guide and an instrument flight hood by Jeppesen—basics for IFR. (Jeppesen-Sanderson photo)

Flying solely by reference to these is what instrument flying is all about. This is a Cessna Skylane. (Cessna photo)

■ **MINIMUM VFR WEATHER CONDITIONS (FAR 91.105)**

Except when operating Special VFR (according to FAR 91.107), or IFR, all aircraft must be operated in accordance with these minimum weather conditions:

Altitude	Flight visibility	Distance from clouds
1,200 feet or less AGL		
—within controlled airspace	3 statute miles	500 ft below 1,000 ft above 2,000 ft horizontally
—outside controlled airspace	1 statute mile*	clear of clouds
More than 1,200 feet AGL, but below 10,000 feet MSL		
—within controlled airspace	3 statute miles	500 ft below 1,000 ft above 2,000 ft horizontally
—outside controlled airspace	1 statute mile	clear of clouds
More than 1,200 feet above the surface, and above 10,000 feet MSL	5 statute miles	1,000 feet below 1,000 feet above 1 mile horizontally

*Helicopters may be operated below 1,200 AGL in visibilities below one mile outside of controlled airspace provided that they fly slowly enough to see air traffic or obstructions in time to avoid a collision.

with at least 5 of these being in an airplane or helicopter, as appropriate. Reading between the lines, this means that you *could* take 5 hours of IFR instruction in a plane from an IFR instructor, 20 hours on a simulator with same, and do 15 hours of instrument practice with a qualified safety pilot, who doesn't have to be an instructor.

In fact, you'll probably need more than 40 hours to get the rating. I used to teach instrument flying, and I know. It's not something you just pick up in a few sessions. It takes *work* and *understanding*. You must be able to fly the airplane on instruments completely automatically, so the work part goes into the navigating and management, not the flying. This takes lots of training and experience. But it's the best safety factor available to a pilot. The discipline and precision IFR training gives you will stand you in good stead for *all* your flying. It'll also probably reduce your insurance rate.

The written exam

A lot of people think the IFR written is tough, but it isn't at all if you train for it. There are plenty of good books on the subject (one of which I wrote for the Canadian market—called the *Fostair Instrument Course*, available from **Aviation Training Systems Ltd., 633 Main Street, Saskatoon, Sask, S7H 0J8, Canada, telephone (306) 653-4656).** The FAA now makes all the exam questions available in advance (there are hundreds of them), and then tests you on a few of them. So the best courses now teach you how to answer all the questions, and when you take the exam it's relatively easy, since you won't have any nasty surprises.

Weekend cram courses

There are several weekend ground-school courses, designed to take you through the exam study in a couple of days, and you take the exam as soon as you're finished. These are well reviewed by their users. Some of them are:

Accelerated Ground Schools, Inc.
Box 43548
Atlanta, GA 30336
(404) 696-2442 in Georgia
(800) 241-4992 elsewhere

AOPA Air Safety Foundation
Box 5800
Washington, DC 20014
(301) 951-3967 in Maryland
(800) 638-0853 elsewhere

Aviation Seminars
Box 294
Princeton Junction, NJ 08550
(609) 799-2120 in New Jersey
(800) 257-9444 elsewhere

Aviation Training Center, Inc.
6635 South Dayton
Englewood, CO 80111
(303) 770-6000 in Colorado
(800) 525-7454 elsewhere

Aviation Training Enterprises, Inc.
DuPage County Airport
Route 64
West Chicago, IL 60185
(312) 584-4700 in Illinois
(800) 323-0808 elsewhere

Study guides

Several publishers claim to offer the complete set of exam questions and answers for home study. Some are:

Acme School of Aeronautics
Terminal Building
Meacham Airport
Fort Worth, TX 76106
(817) 625-4336

Instrument Rating Answer Book, by Aviation Test Prep., available from **Aviation Book Company, 1460 Victory Boulevard, Glendale, CA 91202, telephone (213) 240-1771.**

ATC Instrument Rating-Airplane Manual (see **Aviation Training Center** address, above).

Instrument Rating Course (Mach 2), Instrument Rating Manual, Instrument Rating Study Pak, all by **Jeppesen-Sanderson, 55 Luverness Drive East, Englewood, CO 80112, telephone (303) 779-5757.**

Books

Here are some of the best books on instrument flying: *Flying IFR* by Richard Collins (Delacorte); *Instrument Flying* by Richard L. Taylor (Macmillan); *Weather Flying* by Robert N. Buck (Macmillan); *The Instrument Rating* (Pan American Navigation Service); *Instrument Flight Manual* by William Kershner (Iowa State University Press).

Flight Simulators

With the ever-increasing cost of fuel and labor, the flight simulator is gaining wide acceptance in the training business. The old original Link trainer is seldom found these days. However, there is a bevy of replacements. The closest thing to the old Link is the new Link—the GAT-1 (GAT stands for General Aviation Trainer). This vaguely resembles the original—it is shaped like a cartoon airplane, with tiny stub wings in some cases, and it revolves and tilts and banks just like its grandfather. You can rent time on a GAT-1 for about $15 an hour. It can be flown solo, so this is quite useful for practice at low cost. **Link Division, The Singer Co., Binghamton, NY 13902, telephone (607) 772-3011.**

Rudy Frasca was a Link instructor in the U.S. Navy, and after many years of tinkering with parts and much philosophizing as to what the machine should

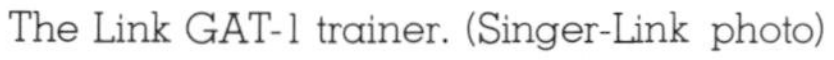
The Link GAT-1 trainer. (Singer-Link photo)

Jeppesen's Mach 2 Instrument Rating Course. (Jeppesen-Sanderson photo)

Frasca Model 101G simulator—a generic light aircraft. (Frasca photo)

Frasca DC 9 simulator. (Frasca photo)

do, he developed the Frasca Simulator. This does not move, but has very sophisticated avionics capabilities, if desired. Rudy started out with fairly simple general-aviation trainers and gradually improved the level of sophistication. Now he's building DC-9 simulators and beyond. **Frasca Aviation, Inc., 606 South Neil Street, Champaign, IL 61820, telephone (217) 359-3951.**

The most ubiquitous ground trainers to be found now are the ATC series table-top models. These are simple-procedures trainers ranging in sophistication from the basic ATC 510, which is no longer in production, to the sophisticated ATC 610 J and K models, which feature more realistic aerodynamic characteristics, fast set-up, an optional remote plotter to record flights, and more. You can pick up a used ATC 510 for a few hundred dollars. The top of the line models cost upwards of $5,000.

ATC has also introduced a helicopter IFR trainer, priced at about $15,000, the ATC 112H.

ATC 610—Everybody's personal simulator. Put one on a table in your basement! (ATC photo)

ATC offers a complete package of software in the form of audiotape cassettes, which contain a variety of realistic IFR assignments based on several areas in the United States—for example, an IFR flight from Monmouth County Airport, Belmar, NJ, to Mercer County Airport, Trenton, NJ, featuring VOR tracking, a VOR hold, and an ILS approach; or a trip from Riverside Airport in California to Ontario International Air-

ATC's new helicopter simulator. (ATC photo)

port, involving ADF tracking, an ADF hold, an ILS approach, a missed approach, a diversion to El Monte Airport, and a back-course localizer approach there. These procedures can be flown solo, so they are useful for practice at any time. **Analog Training Computers, Inc., 185 Monmouth Parkway, West Long Branch, NJ 07764, telephone (201) 870-9200.**

Aviation Simulation Technology (Hanscom Field-East, Bedford, MA 01730, telephone (617) 274-6600) has designed a simulator based on the Mooney 201, and they now have a twin model—the 300. Both are equipped with the latest King Silver Crown digital avionics. The twin model, which I flew recently, has realistic engine noise—it even presents prop-synch problems—and can be used for practicing engine-out procedures. It also features fast set-up and a remote plotter.

Flightmatic, Inc. (150 Riser Road, Teterboro, NJ 07608, telephone (201) 933-5134) produces a variety of trainers covering aircraft ranging from the Cessna 150 at about $20,000 to the Cessna Golden Eagle at about $72,000. One device that can be attached to any current Flightmatic model is the Co-Nav, which costs about $12,000. By means of latitude and longitude inputs, this can create any geographical area in the world, thus enabling a pilot to practice in an unfamiliar area in advance of a trip.

Pacer Systems, Inc. (87 Second Avenue, Burlington, MA 01803, telephone (617) 272-5995) offers several simulators and cockpit procedures trainers, starting with a table-top model at about $5,000, ranging through the recently introduced Gulfstream American Tiger simulator, to Twin Otter and Cessna Citation procedures trainers. All Pacer trainers can be programmed to duplicate any area.

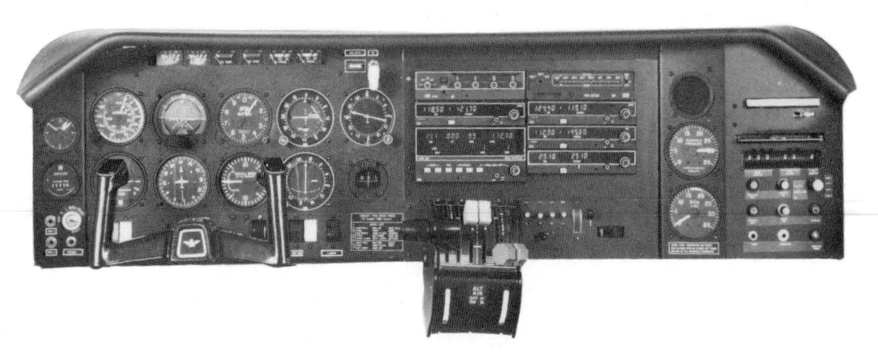

AST's twin simulator, the Model 300. (AST/Hutchins photo)

The Pacer Mark III simulator. (Pacer/Phototech photo)

FlightSafety International, Inc. (Marine Air Terminal, La Guardia Airport, Flushing, NY 11371, telephone (212) 476-5700) builds simulators for advanced aircraft, and they operate training centers throughout the United States, having the largest civilian-owned

The Learjet simulators at FlightSafety's Wichita Learning Center. (FlightSafety International Photo)

On final approach in the FlightSafety Learjet 36 simulator with full visual display. (FlightSafety International photo)

fleet of simulators. They specialize in business jets and turboprops, as well as several airliners. These are the most sophisticated simulators available, with motion and computerized visual terrain depiction in many cases. The following aircraft are simulated, or will be soon:

BA HS 125
BA HS 125-700
Beech King Air 90/100 and 200
Bell 222 helicopter
Canadair Challenger
Cessna Citations I, II and III
Cessna Conquest 421
Convair 340
Dassault Falcon 10, 20 and 50
Douglas DC 6
Douglas DC 10
Gates Learjet 20s, 30s, and 50s
Gulfstream I, II and III
IAI Westwind
Jet Commander
Lockheed Electra
Lockheed Jetstar
Mitsubishi MU 2
Piper Cheyenne I, II, and III
Piper Navajo
Rockwell International Sabreliner
Swearingen Merlin III

. . . and through an affiliate in France . . .

Aerospatiale/BA Concorde
Airbus Industrie A 300

FlightSafety, by the way, now has a subsidiary—Marine Safety—and they offer simulators for supertankers and liquid natural gas tanker vessels. These are the only simulators located at La Guardia Airport. All aircraft simulators are located elsewhere in the United States, wherever they are needed.

Flying Clubs

A flying club is a good way for a group of people to own and operate one or more airplanes. Some clubs are made up of half a dozen folks and an old Piper Tri Pacer, others have hundreds of members and a fleet of ten or more aircraft of different types. Some flying clubs are essentially non-profit cooperatives that exist to enable their members to fly good equipment relatively inexpensively. Others are commercial enterprises. According to the Aircraft Owners and Pilots Association, about 40,000 pilots belong to about 3,000 flying clubs operating about 4,500 aircraft.

AOPA **(Box 5800, Washington, DC 20014)** publishes a booklet "The Flying Club," which outlines the whole picture, from organizational structures to operations, from financing and insurance to problem areas. The FAA offers an Advisory Circular, #00-25, "Forming and Operating a Flying Club," available free from the **Department of Transportation, Publications Section, M-443.1, Washington, DC 20590.**

How to find a flying club

Flying clubs can be elusive. There are no federal registration or organizational requirements for flying clubs, *per se*. Clubs are operated under FAR Part 91, just like private aircraft. Ask at your local airport if there are any clubs based there. If you look at the notice board, you'll often find invitations for membership. Or look at the classified ads in *Flying* or *AOPA Pilot*. Another source is your local FAA General Aviation District Office. If you speak to a General Aviation Operations Inspector, you might be able to find out about local clubs, but you'll get a far from comprehensive list. If you have a state aeronautics commission, it may be able to help, too.

Beech Aero Clubs

Beechcraft has Beech Aero Clubs associated with their Aero Centers. This is the closest thing to a national club, offering interclub privileges and the like. Members get reduced flying rates at most Beech Aero Clubs, versus the rates charged by the local school. They also have an attractive clubroom and a newsletter called *BAC Talk*. Each year there is the "Beech Aero Club Roundup," a national fly-in held at various resorts. There are over 5,000 members in about 100 clubs. Contact **Beech Aircraft (Wichita, KS 67201, telephone (316) 681-7059)** for their "Aero Club Adventure Kit," which lists the locations of each club and describes the facilities in greater detail.

Aerobatics

Have you ever looped the loop? It's quite all right, really. If it's done right, you'll have positive "G" all the way round. The earth and sky will change places while you feel a little heavy, and then they'll change places again, and that's it! Theoretically, *any* airplane can be looped, but to do so in one that is not licensed for aerobatics is quite illegal. In Europe, they take these things much more seriously, and most

The Christen Eagle. One of the great aerobatic aircraft. (Christen Industries photo)

production aircraft are aerobatic. In the United States, only a few are. Many homebuilt aircraft are specifically designed for aerobatics, notably the Christen Eagle, which is available in kit form **(Christen Industries, Inc., 1048 Santa Ana Valley Road, Hollister, CA 95023, telephone (408) 637-7405).**

Production aerobatic aircraft include the Bellanca Citabria and Decathlon derivations of the old Aeronca Champion **(Bellanca Aircraft Corporation, Box 69, Alexandria, MN 56308, telephone (612) 762-1501).** The word Citabria is in fact "airbatic" spelled backwards (I suppose "Citaborea" sounds too much like a disease). These aircraft are approved for most maneuvers, with the Decathlon having the Christen inverted-fuel and -oil systems, which permit extended inverted flight. The Decathlon also features a symmetrical airfoil section for improved inverted flight performance. There are three Decathlon models: the 8KCAB, with a 150-hp engine and a fixed-pitch prop; the CS 8KCAB, which is the same airplane but with a constant-speed prop; and the Super Decathlon 8KCAB-180, which has a 180-hp engine driving a constant-speed prop. The Decathlons are stressed for +6 and −5 Gs.

Cessna Aircraft Company (Wichita, KS 67201, telephone (316) 685-9111) builds the Cessna 152 Aerobat, which is the basic 152 with some extra beefing up where needed to handle the stresses of aerobatics. It can handle +6 and −3 Gs , which allows it to handle rolls, spins, chandelles, loops, and other relatively mild aerobatics.

The Cessna Aerobat can handle the simple stuff—but stay away from the heavier maneuvers! (Cessna photo)

Beech used to offer an aerobatic straight-tail Bonanza—in fact I think you can still order one. They also had an aerobatic Musketeer a few years ago. Their new Skipper is approved for spins, but not for other aerobatics. Likewise the Piper Tomahawk may be spun, but not looped, etc.

Pitts Aerobatics (Box 547, Afton, WY 83110, telephone (307) 886-3151) offers the Pitts S-1S Special as a factory-built aircraft (for about $32,000) or as a kit for the do-it-yourselfer. This single-seater biplane is a favorite aerobatic machine, stressed for all types of maneuvers, including tail-slides, lomcovaks, snaps, and inverted maneuvers. It is powered by a 180-hp Lycoming. There is also a two-seat version, the S-2A, with a 200-hp Lycoming, and a single-seat version of the S-2A called the S-2S, with a 260-hp engine.

Bellanca Decathlon. This photo is printed right side up (note shadows)! (Bellanca photo)

Beech aerobatic Bonanza shows off. (Beech photo)

Marion Cole hard at work. (Marion Cole photo)

Many display aerobatic pilots use Pitts aircraft. Equivalent home-built versions using factory-made kits are called S-1E and S-2E, respectively, while the version built from plans is called the S-1D.

Plenty of ex-military aircraft are around that are legal for aerobatics. Some of these include the Beech T-34 Mentor, the North American T-6 and T-28, and the Stearman. There are others. Aerobatics training, of course, is a part of every military pilot's background.

Aerobatic instruction

There are several schools and instructors around that offer aerobatic courses, using Bellancas, Cessna Aerobats, and Pitts Specials. **Marion Cole (6615 North Park Drive, Shreveport, LA 71107, telephone (318) 929-2618)** offers a special course in his *Bonanza* for Bonanza and other high-performance aircraft drivers. This consists of five hours in the air plus extensive ground briefings, and covers loops, rolls, barrel rolls, spins, chandelles, and Cuban eights. The course takes about three days to complete and costs $650. Cole's aircraft is based at Shreveport Downtown Airport, and he needs several weeks' notice to get you in his schedule. Marion also flies airshows about a dozen times a year, charging $750 for a 15-minute show within a 250-mile radius of his home base. His performance ends with an inverted ribbon pick-up.

Other aerobatic courses are offered by:

Aero Sport
Box 1989
St. Augustine, FL 32084
(904) 824-6230
(Pitts, Great Lakes, Citabria)

Berwick Airport, PA 18603
(717) 759-2411
(Pitts)

Jim Holland Aerobatic School
Pompano Aviation, Inc.
1006 NE 11 Avenue
Pompano Beach, FL 33060
(305) 946-1136
(Citabria, Decathlon, Pitts)

Bill Thomas
14532 SW 129 Street
Miami, FL 33186
(305) 253-7187
(Decathlon, Pitts)

Williams Aviation, Inc.
Springfield, TN 37172
(615) 384-2526
(Decathlon, Pitts)

There are many more throughout the country.

Bellanca Aircraft has developed a *Pilot Proficiency Program*, which is available at some Bellanca dealers. It consists of a basic introduction to such maneuvers as loops, rolls, and spins, done in Citabria and Decathlon aircraft. However, it is not an aerobatic course. It was developed by Debbie Gary, the well-known aerobatic aviatrix, who zooms a Bellanca Viking around the sky at air shows.

Associations

The International Aerobatic Club, a division of the Experimental Aircraft Association **(Box 229, Hales Corners, WI 53130, telephone (414) 425-4860),** publishes a magazine, *Sport Aerobatics*. You must be a member of EAA to join the IAC. **The Aerobatic Club of America (Box 401, Roanoke, TX 76262, telephone (817) 430-1459)** publishes a monthly *ACA Newsletter*. It sponsors the National Aerobatic Championships.

Books

Among books on the subject of aerobatics are: *Aerobatics*, by Neil Williams (Doubleday); *Aerobatics*, by C. R. O'Dell (St. Martins); *Modern Aerobatics and Precision Flying*, by Harold Krier and Bill Sweet (Modern Aircraft); *Primary Aerobatic Flight Training* by A.C. Medore (Aviation Book Co.).

The lure of water flying. This is a Cessna Hawk XP. (Cessna photo)

Water Flying

Water flying is big fun. The first time I ever flew a seaplane, I automatically levelled off at 200 feet after takeoff. It seemed like a perfectly natural altitude. When you're flying over water in a seaplane, you feel quite safe, because if the engine should quit on you all you have to do is put it down right where you are. One time I was flying a Volmer Sportsman amphibian off Paugh Lake in Ontario, Canada, and in an hour I did 30 takeoffs and landings—just took off, climbed a few feet, did a little turn, and landed, then off again, up, and down again. With a whole lake at your disposal, you have an enormous airport, and you can do virtually anything you want (always respecting the boats, of course).

An amphibian will land on water or land, while a pure seaplane will only land on water. There are two configurations of water-landing airplane—floatplanes and flying boats. Floatplanes are usually landplanes with pairs of enormous pontoons hanging down, while flying boats are designed from scratch to sit in the water when on the surface of mother earth. Floatplanes offer more flexibility in docking and mooring, since they usually fit at a conventional boat dock. Flying boats are usually too low in the water to do that, and so they often need special docking facilities. That's why most bush planes are high-wing monoplanes with floats, such as the DHC Beaver and Otter, the Norduuyn Norseman, the Cessna 180 and 185, and such. You don't see too many Lake Amphibians and Seabees in heavy bush service.

Almost all flying boats are amphibians, but this is not true of floatplanes, which come either way. Of course, it is not a very big deal to get the floats on and off—a few hours will do it, so you can change your aircraft

One of the heavies, Cessna's Turbo Stationair 6 on amphib floats. (Cessna photo)

over from landplane to seaplane and back to suit the season. Amphibious floats are very expensive—almost the cost of a small airplane—so their use must be justified by need or extreme wealth.

The **Seaplane Pilots Association (Box 30091, Washington, DC 20014, telephone (301) 951-3895)** offers much help to the current or potential seaplane pilot. Membership costs $18 a year and includes a copy of the *Water Flying Annual*, which has much useful data and interesting anecdotes about splash flying. To get a seaplane rating, you must pass a flight test appropriate to the type of certificate you hold and show the FAA that you have received instruction from an authorized flight instructor. There is no written-test or minimum-experience requirement. It usually takes about 5 to 10 hours to get a rating. Most of the training relates to maneuvering the seaplane on the water and, of course, takeoffs and landings. There isn't too much difference in handling between a seaplane and a conventional aircraft when they are in the air.

A useful book that answers most questions about seaplane flying is *How to Fly Floats*, by Jay Frey, published by Edo-Aire, the floatmakers. It costs $2.00, and may be ordered from **Edo-Aire Seaplane Division, 65 Marcus Drive, Melville, NY 11746 telephone (516) 293-4000.**

Other float firms are:

Aqua Floats by Capre
805 Geiger Road
Zephyrhills, FL 33599
(813) 782-9541

Canadian Aircraft Products Ltd.
2611 Viscount Way
Richmond, BC V6V 1M9, Canada

Fiberfloat Corp.
895 East Gay Street
Bartow, FL 33830
(813) 533-8001

DeVore Aviation Corp. (PK Floats)
6104 Kircher Boulevard NE
Albuquerque, NM 87109
(505) 345-8713

Wipline, Inc.
South Doane Trail
Inver Grove Heights, MN 55075
(612) 451-1205

The only current production amphibian is the Lake Buccaneer, a four-seater with a 200-hp engine. It cruises at 150 mph and costs about $60,000 equipped. The Ca-

A brace of Buccaneers by Lake. (Lake photo)

nadians have been working away on the Trigull, a sort of 1980s derivative of the Seabee, and it is reported to be going into production with technical and marketing assistance from Grumman, who were famous for their twin amphibs a few years ago. **Trident Aircraft, Ltd., Box 2428, Victoria International Airport, Sydney, BC V8L 3Y4, Canada, telephone (604) 656-7294.** The Teal amphib has been on-again off-again in its availability. Recently, it was off, pending financing.

Although the Seabee is no longer in production, there are a few of them around. A twin conversion is available from **Stol Aircraft Corporation, Norwood Metropolitan Airport, Norwood, MA 02062, telephone (617) 769-4030.** This puts two 180-hp Lycomings on the wings, adds a bit of length to the span and the fuselage, and adds a lot of performance to the airframe.

There are at least four homebuilt amphibians available in plan form. The original is the Volmer VJ-22 Sportsman, designed by Volmer Jensen. I have flown the original prototype, and I also did the initial test-flying on a homebuilt one a few years ago. This is a two-place aircraft that uses the wings from an Aeronca Champion or equivalent. The original design called for a pusher engine installation, and the one I was involved in started out that way, but was converted to a tractor installation of 125 hp after a year, which improved performance considerably. **Volmer Aircraft, Box 5222, Glendale, CA 91201, telephone (213) 247-8718.** Another design is the Osprey 2, also a two-seater. This is a very sleek aircraft, designed by George Pereira. **Osprey Aircraft, 3741 El Ricon Way, Sacramento, CA 95825, telephone (916) 482-0340.** The Taylor Coot is another two-place amphibian, designed by **Molt Taylor, Box 1171, Longview, WA 98632, telephone (206) 425-9874.**

The Osprey 2 homebuilt amphibian shows its sleek lines. (Osprey photo)

The prototype Volmer Jensen VJ-22 Sportsman homebuilt amphibian. (Volmer Aircraft photo)

The Taylor Coot homebuilt kisses the water. (Molt Taylor photo)

If you ever see what you think is a tricycle-gear Seabee and you think it's a Lane Riviera and it isn't, it's probably a Spencer Amphibian Aircar. It was designed by the originator of the Seabee and is available in kit and plan form from **Spencer Amphibian Aircar, 8725 Oland Avenue, Sun Valley, CA 91352, telephone (213) 767-7042.**

Not all production aircraft can take floats, but a surprising number can. A supplemental-type certificate is required from the FAA for a float installation. Among aircraft available on floats are these:

- Beech 18
- Bellanca Citabria
- Bellanca Scout
- Cessna 150/152
- Cessna 170/172
- Cessna 180
- Cessna 185
- Cessna 206
- Cessna Hawk XP
- De Havilland Beaver
- De Havilland Otter
- De Havilland Twin Otter
- Helio Super Courier
- Interstate Arctic Tern
- Maule Lunar Rocket
- Maule M-5
- Piper Cherokee 180
- Piper Cherokee Six
- Piper J 3 Cub
- Piper Pacer
- Piper Super Cruiser
- Piper Super Cub
- Piper Tri-Pacer
- Taylorcraft F 19

Gliding

A private sailplane spotted at Sky Manor, NJ. This is a Pilatus B 4. (Tim Foster photo)

To soar like a bird . . . to drift on the wind . . . to travel almost silently on long, thin wings . . . this is the promise of gliding, or more correctly, soaring. When I learned to fly, we derisively referred to gliders as "horseless airplanes," but that was before I actually flew in one. The last time I did so was at El Mirage, in the Mojave Desert of California. My reaction was that gliding was like a beautiful, sensual massage—relaxing and exciting. We were towed aloft by a Bellanca Scout and released at 3,000 feet AGL. We flew a lazy pattern around the airport—there wasn't much lift around—and came in for a landing a few minutes later. Pure fun!

The Soaring Society of America (Box 66071, Los Angeles, CA 90066, telephone (213) 390-4448) is the prime mover in soaring in the United States. It is the national representative of the Federation Aeronautique Internationale for soaring activities. The FAI awards badges to soaring pilots who have accomplished certain activities. These are awarded through the SSA in the United States. The requirements are as follows:

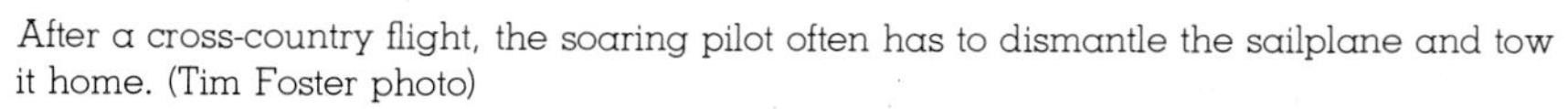
After a cross-country flight, the soaring pilot often has to dismantle the sailplane and tow it home. (Tim Foster photo)

Badge: Requirements

A: Knowledge of glider nomenclature and handling procedures, airport rules, FARs, tow equipment signals and procedures, hold FAA student glider pilot certificate, and have made a solo flight.

B: Solo flight of at least 5 minutes duration above a point of release, or 30 minutes duration after release from a 2,000-foot tow.

C: Dual soaring practice; knowledge of cross-country soaring procedure, glider assembly and retrieving; dangers of cross-country flying; demonstrated ability to carry out simulated cross-country landing in restricted areas without reference to an altimeter; solo flight of at least 30 minutes duration above point of release, or 60 minutes duration after release from a 2,000-foot tow.

Silver C: Straight-line flight of at least 50 km (31.3 sm); remain aloft at least 5 hours; gain at least 1,000 meters (3,281 feet) in altitude.

Gold C: Straight-line flight of at least 300 km (186.4 sm); remain aloft at least 5 hours; gain at least 3,000 meters (9,842 feet) in altitude.

Diamond: (Filling each of these requirements earns one diamond, for a maximum of three): Straight-line flight of at least 500 km (310.7 sm); triangular out-and-return flight of at least 300 km (186.4 sm); gain of at least 5,000 meters (16,404 feet) of altitude.

The FAA licenses glider pilots just as it does power pilots. There is no medical-exam requirement—just an affirmation that there are no known medical deficiencies that would prevent you from flying a glider. To hold a student certificate, you must be at least 14 years old. A private certificate requires you to be at least 16, and a commercial at least 18. There are written and flight tests for the private, commercial, and instructor ratings. To obtain a private certificate, you must have at least 7 hours of solo glider time, including 20 flights if by aero tow; for a commercial, you must have at least 25 hours of flight time, including 20 in gliders, 100 glider flights as pilot in command, or 200 hours of pilot time in heavier-than-air aircraft with 20 glider flights as pilot in command; for an instructor rating you need a commercial certificate, a written, and a flight test.

Where to go soaring

The Soaring Society of America publishes a *Directory of United States Soaring Sites and Organizations*, available for 50¢ from them (see address above). **Soaring Centers of America (Schweizer Aircraft Corporation, Box 147, Elmira, NY 14902, telephone (607) 739-3821)** offers a *Soaring Portfolio* for $7.95 that includes a glider-pilot log book, a copy of *Soaring* magazine, a book called *Start Soaring*, and a list of the "Soaring Centers of America," which is the Schweizer dealer network.

Schweizer is the main sailplane builder in the United States. They operate the Schweizer Soaring School from May through September at Elmira, with holiday courses that will teach you to be a glider pilot on your vacation.

Sailplanes

Schweizer sells the model 2-33 two-seater sailplane, which weighs 600 pounds empty and 1,040 fully loaded. Solo it stalls at 31 mph and cruises at 45 mph, with a glide ratio of 22.25:1. Dual it stalls at 35 mph and cruises at 52 mph with the same glide ratio. The 2-33 features all-metal wings and a tubular-steel, fabric-covered fuselage. The model 1-26E is

Schweizer's basic trainer—the 2-33. (Schweizer photo)

Schweizer 1-26 basic sailplane. (Schweizer photo)

Schweizer 1-35 high performance sailplane. (Schweizer photo)

a single-seater that weighs 445 pounds empty and 700 loaded. It stalls at 33 mph and cruises at 53 mph with a 23:1 glide ratio. The new model 1-35 is a high-performance sailplane with a 15-meter (49 feet) span. It weighs 660 pounds loaded but can take on an additional 320 pounds of water ballast for certain flight requirements. It stalls at 32 mph, and has a glide ratio of 40:1.

Laister Sailplane Products (2712 Chico Avenue, South El Monte, CA 91733, telephone (213) 442-4945) builds the Nugget high-performance sailplane with a 15-meter span. It has a glide ratio of 38:1 at 57 mph and weighs 900 pounds, including 175 pounds of water ballast, if needed.

All other major sailplanes are built abroad. Most of these are represented in the United States either by **Aerosport (32301 Corydon Road, Elsinore, CA 92330, telephone (714) 674-1584)** or by **Graham Thomson Ltd. (3200 Airport Avenue, Santa Monica, CA 90405, telephone (213) 390-8654),** who offer a catalog *For The Soaring Pilot* free of charge.

Books

Among books available on soaring are *The Complete Beginners Guide to Soaring and Hang Gliding*, by Norman Richards (Doubleday); *The Art and Technique of Soaring*, by Richard A. Wolters (McGraw-Hill); *The Complete Soaring Pilots Handbook*, by Ann Welch, Lorne Welch and Frank Irving (McKay).

Hang Gliding

Hang gliding is one of the fastest-growing aeronautical sports. It is also one of the cheapest and has been one of the most dangerous. The fatality figures in the early years were appalling. However, more experience is entering the business, and the safety level has improved considerably.

Hang gliding was really started in 1891 by Otto Lilienthal at Dervitz, Germany. He made thousands of flights but died in 1896 as the result of a crash. Others pursued his techniques, including Octave Chanute, an American, who perfected a hang glider in 1896, and the Wright Brothers, who started flying in 1899. The Wrights, of course, developed their gliders into a powered airplane, the Wright Flyer, which first flew on December 17, 1903.

Francis Rogallo, a space engi-

The Volmer VJ24 Sunfun. (Volmer Jensen photo)

neer, in 1951 developed a wing for NASA to use as a spacecraft-recovery device, replacing the parachute. Because of stowage difficulties, the wing was not used for this purpose, but extensive flight testing proved it to have excellent flight properties. Many hang gliders today employ the Rogallo wing, which is easy to fly and inexpensive. The pilot hangs beneath the wing. Takeoff is accomplished by running into the wind down a slope. By body movements, the glider's center of gravity is changed, and the pilot can control the craft.

The Volmer VJ11 biplane hang glider. (Volmer Jensen photo)

Other hang gliders use rigid wings and are either monoplanes or biplanes. These are more complicated to fly than the Rogallo type, but they have a much better glide ratio, so they can stay up in the air longer.

The latest trend is toward powered hang gliders—they use tiny gas engines. Some use chain saw motors. These are referred to as "foot-launched" aircraft. I hear that the inimitable Burt Rutan is working on a wind-up, *clockwork*-powered hang glider! Take *that*, energy crisis! Next we'll be seeing rubber-band power, just like the little model planes built in our youth.

The **United States Hang Gliding Association (Box 66306, Los Angeles, CA 90066, telephone (213) 390-3065)** is the main organization for hang-gliding people. There is also the **Hang Glider Manufacturers Association (Box 2711, Van Nuys, CA 91408, telephone (213) 361-8651,** whose members build and sell hang gliders. Here is the most recent list of their members:

Chuck's Glider Supplies
4252 Pearl Road
Cleveland, OH 44109
(216) 398-5272

Delta Wing Kites
13620 Saticoy Street
Van Nuys, CA 91408
(213) 787-6600

Eipper-Formance, Inc.
1070 Linda Vista Drive
San Marcos, CA 92069
(213) 328-9100

Electra Flyer Corp.
700 Comanche NE
Albuquerque, NM 87107
(505) 344-3444

Flight Designs
Box 1053
Salinas, CA 93902
(408) 758-6896

Highster Aircraft
1508 6th Street
Berkeley, CA 94701
(415) 527-1324

JL Enterprises
1150 Gold Country Road
Belmont, CA 94002
(415) 592-3613

Manta Products, Inc.
1647 East 14th Street
Oakland, CA 94606
(415) 536-1500

Seagull Aircraft
1160 Mark Avenue
Carpenteria, CA 93013
(805) 684-8331

Seedwings
1096 Via Regina
Santa Barbara, CA 93111
(805) 682-4250

Sky Sports, Inc.
394 Somers Road
Ellington, CT 06029
(203) 872-7317

Sunbird Ultra Light Gliders
12501 Gladstone Avenue, #A 4
Sylmat, CA 91342
(213) 361-8651

Ultimate High
14328 Lolin Lane
Poway, CA 92064
(714) 748-1739

Ultralight Products
Box 582
Rancho Temacula, CA 92390
(714) 676-5652

Wills Wings, Inc.
1208 H East Walnut
Santa Ana, CA 92701
(714) 547-1344

A typical Rogallo-type hang glider is the Sunbird Nova, which sells for about $1,250. The copy advertising it sounds a little strange to the uninitiated: "Sunbird Ultralight Gliders introduces the NOVA, a 'state-of-the-art' soaring ship with 130° nose angle, *no deflexers* and a ten meter span." Hang gliders come in different sizes, according to pilot's weight. For example, the Nova comes in four models as follows:

■ MODEL NUMBER	150	170	190	230
Wingspan (feet)	29.25	31.0	32.9	36.0
Area (square feet)	150	170	190	230
Pilot weight pounds	100–140	120–160	140–190	175–220

■ PERFORMANCE FIGURES ARE:

Stall speed	14	mph
Top speed	40	mph
Sink rate	190	fpm minimum
Rate of roll	39°	per second

One person who has been involved in a variety of aeronautical enterprises is Volmer Jensen. He designed the Volmer VJ 22 Sportsman, a two-seater home-built amphibian, and lately has been more active in the foot-launched area of flight. He still offers plans for the VJ 11, a hang glider he designed in 1940. His current designs are the VJ 23 and 24 hang gliders, which can be motorized and are then called the VJ 23E and 24E. Jensen's designs are available only in plan form for building at home. Plans for the Sunfun VJ 24 cost about $55, and the engine installation plans cost an extra $75. The Sunfun with an engine can take off on a level surface, and it can actually climb back to the starting point instead of having to be schlepped back up hill, as is normal, making it useful for more frequent flights. **Volmer Aircraft, Box 5222, Glendale, CA 91201, telephone (213) 247-8718.**

Neither regular nor motorized hang gliders require an FAA aircraft certificate of airworthiness or a pilot's license to be flown.

Volmer VJ 24E Sunfun, a powered hang glider. (Volmer Aircraft photo)

The United States Hang Gliding Association certifies instructors for their members. You can take lessons for about $20 an hour or take an introductory package for about $50. Look in the **Yellow Pages** under **"Gliders—Hang"** for a nearby location.

Books

There are many books on the market about hang gliding. Here is a partial listing:

- *The Complete Beginners Guide to Soaring and Hang Gliding* by Norman Richards (Doubleday)
- *Hang Gliding* by Ross R. Olney (Putnam)
- *Hang Gliding* by Dorothy Schmitz (Crestwood)
- *Hang Gliding: Basic Handbook of Skysurfing* by Dan Poynter (Parachuting Publications)
- *Hang Glider's Bible* by Michael A. Markowski (Tab)
- *The Hang Gliding Book* by William Bixby (McKay)
- *Hang Gliding Handbook: Fly Like a Bird* by George Siposs (Tab)
- *Hang Gliding: The Flyingest Flying* by Don Dedera (Northland)
- *Hang Gliding: Rapture of the Heights* by Lorraine M. Doyle (Aviation Books)
- *Hang Gliding: Riding the Wind* by Otto Penzler (Troll)

- *Hang Gliding and Soaring: a Complete Introduction to the Newest Way to Fly* by James E. Mrazek (St. Martin)
- *Hang Flight: Flight Instruction Manual for Beginner and Intermediate Pilots* by Joe Adleson and Bill Williams (Eco-Nautics)

The Soaring Society of America, Box 6601, Los Angeles CA 90069, telephone (213) 390-4448, offers a beginners kit on hang gliding.

Ballooning

Two brothers named Etienne and Joseph Montgolfier successfully launched the first balloon—a hot air model—on June 4, 1783. Soon after, a physicist named J.A.C. Charles introduced and flew a *hydrogen* balloon, which made an unmanned flight on August 27. As if in a duel, the Montgolfiers responded by sending up in their balloon an animal payload—a cock, a duck, and a sheep—on September 19. And the brothers capped it all by launching the first people to fly in a balloon—Pilatre de Rozier and the Marquis d'Arlandes, who soared aloft on November 21. Charles then flew in a hydrogen balloon on December 1. 1783 was quite a year for flying firsts!

The Montgolfiers burned a mixture of chopped straw and wool to heat the air, believing that the fuel, not the hot air, was what gave the balloon its lifting powers. The alternative method—filling the balloon with explosive hydrogen—was dangerous. Later balloonists used helium, an inert gas with only half the lifting power of hydrogen. Nowadays, the most popular method of levitation for balloonists is hot air created by a propane burner.

Ballooning is booming. Hardly a public event goes by without at least one hot-air balloon lifting off—often as an advertisement. Quite a few balloons have been created that take on the physical characteristics of their sponsor's product.

You can buy a very nice balloon from **Raven Industries, Box 1007, Sioux Falls, SD 57101, telephone (605) 336-2750.** The costs of Raven balloons range from about $7,500 for the most basic 56,400-cubic-foot model to about $13,000 for a 140,000-cubic-foot system. All kinds of optional extras are available, including a suede-edged wicker basket with a padded leather seat and a padded console that will keep two bottles of champagne cool. The champagne is to share with and placate the surprised owner of your landing spot!

Other balloon manufacturers include:

Avian Hot Air Balloon Company
Ridgewood Drive
Spokane, WA 99026
(509) 928-6847

The Balloon Works
RFD 2
Statesville, NC 28667
(704) 873-0503

Cameron Balloons Ltd.
3600 Elizabeth Road
Ann Arbor, MI 48103
(313) 995-0111

Mike Adams Balloon Loft, Inc.
Box 12168
Atlanta, GA 30355
(404) 261-5818

Piccard Balloons
Box 1902
Newport Beach, CA 92663
(714) 642-3545

Semco Balloon
Route 3, Box 514
Griffin, GA 30223
(404) 228-4005

Thunder Pacific
114 Sandalwood Court
Santa Rosa, CA 95401
(707) 546-7124

Balloonist licensing

The free-balloon pilot's certificate may be obtained by anyone in reasonable health—no medical examination is required. A student certificate is needed before starting training. Student pilots must be at least 14 years old and must demonstrate knowledge of Part 91 of the Federal Aviation Regulations (FARs), as well as demonstrate proficiency at preflight preparation, operation of the controls, lift-off and climb, descent and landing, and emergency situations. Students are allowed to fly balloons only under the supervision of a qualified instructor, and they may not carry passengers or fly a balloon for hire.

The private pilot's certificate can be earned by a person who is at least 16 years old and who passes a written exam about FARs, use of navigation charts, recognition of weather conditions and use of weather reports, and the operation of balloons. The person must have received instruction in ground handling and inflation, preflight checks, takeoffs and ascents, descents and landings, and emergency procedures. He or she must have at least 10 hours of flight experience in free balloons, with at least 6 flights under the supervision of an instructor. These flights must include at least the following: two flights of at least 30 minutes duration, one ascent to 3,000 feet above the takeoff point, and one solo flight. (These requirements are for hot-air balloons. Gas balloons have slightly different altitude requirements.)

To get a commercial balloonist's certificate, you must be at least 18 years old, pass a more advanced written test, receive advanced training, have at least 35 hours of flight time (20 of these hours must be in balloons; the rest may be in other aircraft). This flight time must include 10 flights in free balloons, with 6 of these under the supervision of an instructor, two solo flights, two flights of at least one hour's duration, and one flight to 5,000 feet above the takeoff point (10,000 feet, if a gas balloon).

Balloon holiday

A nice way to learn ballooning is to spend a vacation at **The Balloon Ranch, Del Norte, CO 81132, telephone (303) 754-2533.** Here you can take balloon lessons, as well as go horseback riding, play tennis, and go rock-climbing, to say nothing of enjoying gourmet dining and hot-tubs!

Balloon operation

There are three components of balloon flight—inflation, flight, and landing. The maximum wind speed for safe flight is about 8 mph, so most flying is done early in the morning or evening. To inflate the balloon, a portable blower-fan is used to push cold air into the balloon envelope. It takes at least three people (preferably five or six) to accomplish this. The inflation process takes about 15 minutes. When the balloon is about half inflated with cold air, the propane burner is ignited and used to heat the air. As the air warms up, the balloon becomes buoyant.

Flight begins when the balloon lifts free of the ground. The burner and a maneuvering vent control the rate of ascent and descent and are also used to maintain level flight. You can't steer a balloon. It goes where the wind blows. However, since the wind blows at different rates and directions at different altitudes, you can make a rough change of direction by moving higher or lower.

A balloon is landed by cruising at a low altitude until a suitable touchdown spot is selected. By controlling the rate of descent, the skilled balloonist can sometimes put the device close to the chosen landing site, but sometimes the balloon lands where it's not wanted—right in the middle of an interstate highway, for example! Once the balloon is down, the deflation port is opened to discharge the air in the balloon. It takes about 5 minutes to deflate a balloon, and another 20 to 30 minutes to pack it away.

Taking a balloon ride

You can probably arrange a balloon ride with your local balloon club. Look in the Yellow Pages under "Balloons," or contact the **Balloon Federation of America (1508 11th Avenue North, Fort Dodge, IA 50501, telephone (515) 573-8265)** for the club nearest you. A ride will cost about $100, and one or two people can be taken. Riding a balloon is an experience you will always remember. There is absolutely no sensation of motion, and, when the burner is not blasting, all is quiet; you can hear activity on the ground quite clearly.

Reference sources

The most comprehensive book on the subject is *The Encyclopedia of Hot Air Balloons*, by Paul Garrison, a large paperback published by Drake. This lists all the balloon clubs in the USA, as well as the annual ballooning events. Naturally, there is a journal—***Ballooning* Magazine (2516 Hiawatha Drive NE, Albuquerque, NM 87112)**, and there is the **Balloon Federation of America,** mentioned previously.

P A R T T W O

Aircraft

MANUFACTURERS
LIGHT AIRCRAFT RANKINGS
TURBOPROPS
JETS
HELICOPTERS
GYROCOPTERS
AGRICULTURAL AVIATION
ANTIQUE AIRPLANES
HOMEBUILT AIRCRAFT
USED AIRCRAFT

The Aerostar 601B—the high performance turbo-charged twin from Piper's Santa Maria Division. (Piper photo)

Manufacturers

The American aircraft industry builds more flying machines than all other countries in the world put together: the best-selling airliners, utility aircraft, business jets, single- and multi-engine general-aviation aircraft, training planes, agricultural planes, helicopters and military aircraft of all types are built here.

In 1979, the US aircraft industry built over 17,000 general-aviation aircraft, with a total value of $2.2 billion. Of these, 3,995 were exported; their value was over $600 million. Total 1979 production consisted of 282 business jets, 637 turboprop business planes, 2,843 twin-engine aircraft and 13,286 single-engine planes.

The big three aircraft builders are Cessna, Piper, and Beech, in that order. **Cessna Aircraft Company (Box 1521, Wichita, KS 67201, telephone (316) 685-9111)** manufactures the most complete range of aircraft of any company in the world. They build two-seater trainers, single-seat agricultural planes, medium- and high-performance singles, the world's only pressurized single, medium- and high-performance twins, cabin-class twins, turboprops, and the best-selling business jets.

In 1979, the industry shared the total market as follows, based on unit shipments:

US General Aviation Manufacturers 1979 Shipments and Sales

Company	Units shipped	Market share (%)
Cessna	8,380	49
Piper	5,253	31
Beech	1,508	9
Bellanca	440	2
Mooney	439	2
Gulfstream American	404	2
Rockwell International	164	†
Gates Learjet	107	†
Lake	96	†
Swearingen	70	†
Maule	66	†
Miscellaneous	121	†
Totals	17,048	100%

† less than 1%
Source: GAMA

Perhaps a fairer way to assess the manufacturers' relative share of the 1979 market is to look at it in terms of dollar sales:

Company	Sales (in thousands of dollars)	Dollar share (in %)	Unit share (in %)
Cessna	755,037	34	49
Piper	418,949	19	31
Beech	402,143	18	9
Gates Learjet	225,993	10	†
Gulfstream American	138,700	6	2
Rockwell International	110,870	5	†
Swearingen	80,024	4	†
Mooney	30,730*	1	2
Ayres	13,789	†	†
Bellanca	12,679	†	2
Lake	4,512	†	†
Maule	2,276	†	†
Totals	2,195,703	100%	100%

*estimated
†less than 1%
Source: GAMA

Piper Aircraft Corporation (Lock Haven, PA 17745, telephone (717) 748-6711) is a subsidiary of the Bangor Punta Corporation. They still build a version of the Piper Cub, the plane that made them famous, and also offer a fairly complete line, including turboprops, but they haven't offered a pure jet to date. They have plants in Florida as well as the original one in Pennsylvania.

The third of the big three is **Beech Aircraft Corporation (Wichita, KS 67201, telephone (316) 681-7111)**, which traditionally, has built the highest-priced aircraft. Beech offers a full line and concentrates heavily on the turboprop twins. The Beech King Air is used by every branch of the United States armed services as well as hundreds of corporations. Beech has marketed jets in the past (the British HS 125, and before that the French MS 760 Paris), but tops out with turboprops now.

The next manufacturer in market share is Gates Learjet, founded by the legendary Bill Lear. The Lear Jet is almost syn-

Coming from Cessna in the near future, the new Clipper 6-seater twin. (Cessna photo)

onymous with the corporate jet. Most lay people think all business jets are Lear Jets, just as they think all private planes are Piper Cubs. Though Gates Learjet attempted to produce a helicopter some years ago, they now build only jets, every one of them beautiful. They are made in Wichita and marketed from Tucson. **Gates Learjet Corporation, Box 7707, Wichita, KS 67277, telephone (316) 722-5640. Gates Learjet Corporation, Aircraft Marketing, Box 11186, Tucson, AZ 85734, telephone (602) 294-4422.**

Gulfstream American Corporation (Box 2206, Savannah, GA 31402, telephone (912) 964-3000) used to be Grumman American, but it was taken over by American Jet Industries, Inc. of Van Nuys, CA. Builders of the Gulfstream business aircraft, they will soon be offering the unique Hustler turboprop/jet hybrid businessliner. The Gulfstream light aircraft line has been sold, and is expected to be relocated to Northern Ireland under new ownership.

Rockwell International (General Aviation Division, 5001 North Rockwell Avenue, Bethany, OK 73008, telephone (405) 789-5000) was once Aero Commander. They now build high-performance twin-engine aircraft. They discontinued building singles in 1979 due to poor sales. The same company also makes the Sabreliner at its **Sabreliner Division, 6161 Aviation Drive, St. Louis, MO 63134, telephone (314) 731-2260.**

Swearingen Aviation Corporation (Box 32486, San Antonio, TX 78284, telephone (512) 824-9421) is a subsidiary of Fairchild Industries. They make the Metro and Merlin turboprop twins.

Mooney Aircraft Corporation (Box 72, Kerrville, TX 78028, telephone (512) 896-6000) is a subsidiary of Republic Steel. They concentrate on high-performance singles, the Models 201 and 231.

Ayres Corporation (Box 3090, Albany, GA 31706, (912) 883-1440) builds the agricultural aircraft formerly offered by Aero Commander, including a large turboprop biplane.

Bellanca Aircraft Corporation (Box 69, Alexandria, MN 56308, telephone (612) 762-1501) should not be confused with Bellanca Aircraft Engineering, Inc. of West Virginia. Bellanca Aircraft manufactures the Citabria and Decathlon aerobatic singles (this line is for sale), and the Aries high-performance single. The Viking has

The Beech Baron 58P, a pressurized, turbo-charged, 6-seater twin. (Beech photo)

The ultra-high-performance Mooney 231. (Mooney photo)

Bellanca is pinning its hopes on the T-250 high performance single. (Bellanca photo)

One of the discontinued Gulfstream American singles—soon to return to production in Northern Ireland. This is the Tiger (Gulfstream American photo)

been discontinued. The other Bellanca is developing a high-performance single of great beauty, but does not yet have the plane in production.

Lake Aircraft (Box 399, Tomball, TX 77375, telephone (713) 376-5421) makes the four-place Buccaneer amphibian, which started out years ago as the Colonial Skimmer and has gradually evolved into its present form.

Maule Aircraft Corporation (Spence Air Base, Moultrie, GA 31768, telephone (912) 985-2045) concentrates on the STOL (short takeoff and landing) four-place M-5 aircraft. You'll see lots of these in Alaska, but not too many at Teterboro, N.J.

So these, then, are the principal United States manufacturers of general-aviation aircraft.

Looming over the horizon is **Lear Avia Corporation (Box 60000, Reno, NV 89506)** another of the late, great Bill Lear's enterprises coming soon with the Lear Fan, a unique twin-turboprop, single-prop business aircraft, to be built in Northern Ireland. And there's always Tony Fox, creator of the diminutive Foxjet, at **Foxjet International, 6701 West 110th Street, Minneapolis, MN 55438, telephone (612) 944-2255.** And the Windecker Eagle is promised in our future, by **Composite Aircraft Corporation, 523 Ridgeview Drive, Florence, KY 41042, telephone (606) 371-7241.** And the Wing Derringer tiny twin is back in production. This two-seat twin is offered by **Wing Aircraft Company, 2550 Skypark Drive, Torrance, CA 90505, telephone (213) 534-3820.**

In addition there are planes built abroad and assembled

A "green" Falcon 50 is greeted on its arrival in the United States by another 50 and the Goodyear "America." (Falcon Jet photo)

here. For example, **Mitsubishi Aircraft International, Inc. (12700 Park Central Drive, Box 57, Dallas, TX 75251, telephone (214) 387-5600)** builds the Mitsubishi MU-2 in Japan and assembles it in Texas.

Falcon Jet Corporation (Teterboro Airport, Teterboro, NJ 07608, telephone (201) 288-5300) builds Falcon 10s, 20s, and 50s in France, and flies them over for completion in Arkansas.

Light Aircraft Rankings

Manufacturers are always making claims about their aircraft—this is the fastest, this is the most efficient, this carries the greatest load, this has the longest range, and so on.

The *Aviator's Catalog* has spared no expense in bringing you the following rankings of current aircraft data. The tables use the manufacturers' own figures, which have been computer-sorted to give you the various rankings.

Current Production Aircraft Ranked by Basic Price (1980)

The prices shown are the "manufacturers' list price" for the aircraft, with no optional equipment.

■ SINGLE-ENGINE AIRCRAFT

Manufacturer	Type	Price
Cessna	152	16,950
Bellanca	Citabria 7ECA	18,750
Piper	Tomahawk	19,060
Beech	Skipper	19,950
Varga	Kachina	22,950
Bellanca	Decathlon	26,400
Piper	Super Cub	27,110
Cessna	Skyhawk	27,250
Piper	Warrior II	27,790
Bellanca	Scout	28,500
Maule	Lunar Rocket	32,045
Piper	Archer II	34,010
Cessna	Hawk XP	35,950
Beech	Sundowner 180	39,530
Piper	Dakota	41,070
Cessna	180	41,975
Cessna	Cutlass RG	43,395
Cessna	Skylane	44,550
Socata	Rallye 235 GT	44,900
Piper	Turbo Dakota	46,580
Piper	Arrow IV	50,590
Cessna	185	51,275
Mooney	201	52,000
Beech	Sierra 200	53,900
Piper	Turbo Arrow IV	55,730
Mooney	Turbo 231	57,775
Cessna	Stationair 6	58,750
Cessna	Skylane RG	58,750
Bellanca	Super Viking	62,500
Lake	Buccaneer	64,400
Cessna	Turbo Skylane RG	65,500
Cessna	Turbo Stationair 6	66,175
Piper	Saratoga	66,700
Bellanca	Aries T-250	66,750
Cessna	Stationair 8	68,450
Piper	Turbo Saratoga	74,900
Cessna	Turbo Stationair 8	76,200
Cessna	Centurion	76,250
Piper	Saratoga SP	80,200
Cessna	Turbo Centurion	84,100
Piper	Turbo Saratoga SP	88,400
Beech	Bonanza F 33A	91,950
Beech	Bonanza V 35B	91,950
Beech	Bonanza A 36	99,250
Beech	Bonanza A 36 TC	111,250
Cessna	Pressurized Centurion	117,300

The Cessna 152—at $16,950 basic, the lowest-cost mass-produced aircraft available. (Cessna photo)

The Cessna Pressurized Centurion costs $117,300 to start. The radar is extra. (Cessna photo)

■ TWIN-ENGINE AIRCRAFT

Manufacturer	Type	Price
Piper	Seminole	86,960
Cessna	Skymaster	104,200
Beech	Duchess 76*	107,000
Piper	Seneca II	112,230
Cessna	Turbo Skymaster	121,000
Wing	Derringer	125,000
Cessna	310	137,600
Beech	Baron B 55	141,500
Piper	Aztec F	143,300
Cessna	Pressurized Skymaster	159,000
Cessna	Turbo 310	160,650
Beech	Baron E 55	173,750
Piper	Turbo Aztec F	179,205
Piper	Aerostar 600 A	186,120
Beech	Baron 58	201,750
Cessna	335	209,950
Piper	Aerostar 601 B	212,270
Beech	Baron 58 TC	227,300
Piper	Navajo C	234,890
Cessna	340 A	235,950
Cessna	402 C	236,950
Piper	Navajo C/R	249,840
Piper	Chieftain	268,920
Beech	Baron 58P	271,500
Piper	Aerostar 601 P	273,250
Cessna	Chancellor	307,470
Beech	Duke	341,700
Cessna	Golden Eagle	357,470
Cessna	Titan	375,000

At $86,960 basic price, the Piper Seminole is the cheapest twin you can buy. (Piper photo)

The mighty Cessna Titan starts at $375,000. (Cessna photo)

Inside the Titan. (Cessna photo)

Current Production Aircraft Ranked by Gross Weight

Some aircraft are rated by ramp weight, *which accounts for taxi fuel. Others are just rated by maximum gross weight allowable for takeoff. The weight shown here is either the takeoff or ramp weight, as applicable.*

■ SINGLE-ENGINE AIRCRAFT

Manufacturer	Type	Weight lbs
Bellanca	Citabria 7ECA	1,650
Piper	Tomahawk	1,670
Cessna	152	1,675
Beech	Skipper	1,680
Piper	Super Cub	1,750
Bellanca	Decathlon	1,800
Varga	Kachina	1,817
Bellanca	Scout	2,150
Maule	Lunar Rocket	2,300
Cessna	Skyhawk	2,307
Piper	Warrior II	2,325
Beech	Sundowner 180	2,455
Piper	Archer II	2,550
Cessna	Hawk XP	2,558
Piper	Arrow IV	2,570

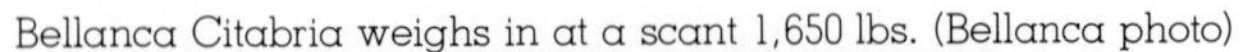
Bellanca Citabria weighs in at a scant 1,650 lbs. (Bellanca photo)

Cessna Pressurized Centurion—a hefty 4,016 lbs. on the ramp, fully loaded. (Cessna photo)

Manufacturer	Type	Weight lbs
Socata	Rallye 235 GT	2,645
Cessna	Cutlass RG	2,658
Lake	Buccaneer	2,690
Mooney	201	2,740
Beech	Sierra 200	2,758
Cessna	180	2,810
Mooney	Turbo 231	2,900
Piper	Turbo Arrow IV	2,900
Piper	Turbo Dakota	2,900
Cessna	Skylane	2,960
Piper	Dakota	3,000
Cessna	Skylane RG	3,112
Cessna	Turbo Skylane RG	3,112
Bellanca	Aries T-250	3,150
Bellanca	Super Viking	3,325
Cessna	185	3,362
Beech	Bonanza F 33A	3,412
Beech	Bonanza V 35B	3,412
Beech	Bonanza A 36	3,612
Cessna	Stationair 6	3,612
Piper	Saratoga	3,615
Piper	Saratoga SP	3,615
Cessna	Turbo Stationair 6	3,616
Piper	Turbo Saratoga	3,617

Manufacturer	Type	Weight lbs
Piper	Turbo Saratoga SP	3,617
Beech	Bonanza A 36 TC	3,666
Cessna	Centurion	3,812
Cessna	Stationair 8	3,812
Cessna	Turbo Stationair 8	3,816
Cessna	Turbo Centurion	4,016
Cessna	Pressurized Centurion	4,016

The Wing Derringer is the lightest twin. (Wing photo)

■ **TWIN-ENGINE AIRCRAFT**

Manufacturer	Type	Weight lbs
Wing	Derringer	3,050
Piper	Seminole	3,800
Beech	Duchess 76	3,916
Piper	Seneca II	4,570
Cessna	Skymaster	4,648
Cessna	Turbo Skymaster	4,652
Cessna	Pressurized Skymaster	4,724

Manufacturer	Type	Weight lbs
Beech	Baron B 55	5,121
Piper	Aztec F	5,200
Piper	Turbo Aztec F	5,200
Beech	Baron E 55	5,324
Beech	Baron 58	5,424
Piper	Aerostar 600 A	5,500
Cessna	310	5,535
Cessna	Turbo 310	5,535
Piper	Aerostar 601 B	6,000
Piper	Aerostar 601 P	6,000
Cessna	335	6,025
Cessna	340	6,025
Beech	Baron 58 TC	6,240
Beech	Baron 58P	6,240
Piper	Navajo C	6,536
Piper	Navajo C/R	6,540
Cessna	Chancellor	6,785
Beech	Duke	6,819
Cessna	402 C	6,885
Piper	Chieftain	7,045
Cessna	Golden Eagle	7,500
Cessna	Titan	8,450

Almost 4¼ tons—the Cessna Titan. (Cessna photo)

Current Production Aircraft Ranked by Useful Load

The useful load here is the maximum useful load available, exluding any optional equipment or fuel. In almost every case, the actual useful load of an aircraft you buy or operate will be lower than these figures indicate, after you have installed radios and other options. However, it's all relative.

■ SINGLE-ENGINE AIRCRAFT

Manufacturer	Type	Useful Load lbs
Cessna	Turbo Centurion	1,795
Cessna	Stationair 8	1,707
Piper	Saratoga	1,695
Cessna	Stationair 6	1,685
Cessna	Centurion	1,679
Cessna	Pressurized Centurion	1,676
Cessna	185	1,674
Cessna	Turbo Stationair 8	1,633
Piper	Saratoga SP	1,629
Piper	Turbo Saratoga	1,617
Cessna	Turbo Stationair 6	1,613
Piper	Turbo Saratoga SP	1,544
Beech	Bonanza A 36	1,421
Beech	Bonanza A 36 TC	1,421

The Cessna Turbo Centurion will carry 1,795 lbs. No wonder it is so popular with the dope smugglers! (Cessna photo)

The Bellanca Decathlon will only lift 540 lbs, but it *is* aerobatic. (Bellanca photo)

Manufacturer	Type	Useful Load lbs
Piper	Dakota	1,398
Cessna	Skylane RG	1,362
Piper	Turbo Dakota	1,321
Cessna	Turbo Skylane RG	1,321
Beech	Bonanza V 35B	1,295
Beech	Bonanza F 33A	1,280
Bellanca	Aries T-250	1,265
Cessna	Skylane	1,254
Piper	Turbo Arrow IV	1,228
Cessna	180	1,166
Bellanca	Super Viking	1,140
Lake	Buccaneer	1,135
Piper	Archer II	1,132
Piper	Arrow IV	1,123
Socata	Rallye 235 GT	1,120
Cessna	Cutlass RG	1,100
Mooney	Turbo 231	1,100
Mooney	201	1,100
Beech	Sierra 200	1,045
Cessna	Hawk XP	1,017

Manufacturer	Type	Useful Load lbs
Piper	Warrior II	985
Beech	Sundowner 180	953
Cessna	Skyhawk	910
Maule	Lunar Rocket	900
Bellanca	Scout	835
Piper	Super Cub	767
Varga	Kachina	692
Bellanca	Citabria 7ECA	583
Beech	Skipper	580
Cessna	152	574
Piper	Tomahawk	562
Bellanca	Decathlon	540

The Cessna Titan will lift 3,634 lbs—43 percent of its gross weight. (Cessna photo)

■ TWIN-ENGINE AIRCRAFT

Manufacturer	Type	Useful Load lbs
Cessna	Titan	3,634
Cessna	Golden Eagle	2,877
Piper	Chieftain	2,824
Cessna	402 C	2,811
Piper	Navajo C	2,533
Beech	Baron 58 TC	2,447

Manufacturer	Type	Useful Load lbs
Piper	Navajo C/R	2,441
Cessna	Chancellor	2,429
Beech	Duke	2,413
Cessna	335	2,276
Beech	Baron 58P	2,241
Cessna	310	2,183
Cessna	340	2,114
Cessna	Turbo 310	2,062
Beech	Baron 58	2,061
Piper	Aerostar 601 B	2,042
Beech	Baron E 55	2,038
Piper	Aztec F	2,017
Piper	Aerostar 601 P	1,944
Beech	Baron B 55	1,888
Piper	Turbo Aztec F	1,878
Cessna	Skymaster	1,861
Piper	Aerostar 600 A	1,763
Cessna	Turbo Skymaster	1,753
Piper	Seneca II	1,729
Cessna	Pressurized Skymaster	1,665
Beech	Duchess 76	1,456
Piper	Seminole	1,440
Wing	Derringer	950

The Wing Derringer will carry 950 lbs. (Wing photo)

Current Production Aircraft Ranked by Seating Capacity*

■ SINGLE-ENGINE AIRCRAFT

Manufacturer	Type	Seats
Cessna	Stationair 8	8
Cessna	Turbo Stationair 8	8
Piper	Saratoga	7
Piper	Saratoga SP	7
Piper	Turbo Saratoga	7
Piper	Turbo Saratoga SP	7
Beech	Bonanza A 36	6
Beech	Bonanza A 36 TC	6
Cessna	180	6
Cessna	185	6
Cessna	Centurion	6
Cessna	Stationair 6	6
Cessna	Turbo Centurion	6
Cessna	Turbo Stationair 6	6
Beech	Bonanza F 33A	5
Beech	Bonanza V 35B	5
Beech	Sierra 200	4
Beech	Sundowner 180	4
Bellanca	Super Viking	4
Bellanca	Aries T-250	4
Cessna	Cutlass RG	4
Cessna	Hawk XP	4
Cessna	Skyhawk	4
Cessna	Skylane	4
Cessna	Skylane RG	4
Cessna	Turbo Skylane RG	4
Lake	Buccaneer	4
Maule	Lunar Rocket	4
Mooney	201	4
Mooney	Turbo 231	4
Piper	Archer II	4
Piper	Arrow IV	4
Piper	Dakota	4
Piper	Turbo Arrow IV	4
Piper	Turbo Dakota	4
Piper	Warrior II	4
Socata	Rallye 235 GT	4
Beech	Skipper	2
Bellanca	Citabria 7ECA	2
Bellanca	Decathlon	2
Bellanca	Scout	2
Cessna	152	2
Piper	Super Cub	2
Piper	Tomahawk	2
Varga	Kachina	2

*including pilot

There are eight seats in the Cessna Stationair 8—the largest capacity single in production. (Cessna photo)

The Piper Super Cub two-seater, still in production after almost 50 years as essentially the same airplane. (Piper photo)

With 10 seats, the Piper Navajo Chieftain is the mainstay of the commuter airline business. (Piper photo)

■ TWIN-ENGINE AIRCRAFT

Manufacturer	Type	Seats
Cessna	Titan	10
Piper	Chieftain	10
Cessna	402 C	8
Cessna	Chancellor	8
Cessna	Golden Eagle	8
Piper	Navajo C	8
Piper	Navajo C/R	8
Piper	Seneca II	7
Beech	Baron 58	6
Beech	Baron B 55	6
Beech	Baron E 55	6
Beech	Duke	6
Beech	Baron 58 TC	6
Beech	Baron 58P	6
Cessna	310	6
Cessna	335	6
Cessna	340	6
Cessna	Skymaster	6
Cessna	Turbo 310	6
Cessna	Turbo Skymaster	6
Piper	Aerostar 600 A	6
Piper	Aerostar 601 B	6
Piper	Aerostar 601 P	6
Piper	Aztec F	6
Piper	Turbo Aztec F	6
Cessna	Pressurized Skymaster	5
Beech	Duchess 76	4
Piper	Seminole	4
Wing	Derringer	2

The Wing Derringer seats two. (Wing photo)

Current Production Aircraft Ranked by Horsepower

In the case of twins, the horsepower shown is the total of both engines.

■ SINGLE-ENGINE AIRCRAFT

Manufacturer	Type	HP
Cessna	152	110
Piper	Tomahawk	112
Beech	Skipper	115
Bellanca	Citabria 7ECA	115
Bellanca	Decathlon	150
Piper	Super Cub	150
Varga	Kachina	150
Cessna	Skyhawk	160
Piper	Warrior II	160
Beech	Sundowner 180	180
Bellanca	Scout	180
Cessna	Cutlass RG	180
Piper	Archer II	180
Cessna	Hawk XP	195
Beech	Sierra 200	200
Lake	Buccaneer	200
Mooney	201	200
Piper	Arrow IV	200
Piper	Turbo Arrow IV	200
Piper	Turbo Dakota	200
Mooney	Turbo 231	210
Cessna	180	230
Cessna	Skylane	230
Cessna	Skylane RG	235
Cessna	Turbo Skylane RG	235
Maule	Lunar Rocket	235
Piper	Dakota	235
Socata	Rallye 235 GT	235
Bellanca	Aries T-250	250
Beech	Bonanza A 36	285
Beech	Bonanza F 33A	285
Beech	Bonanza V 35B	285
Beech	Bonanza A 36 TC	300
Bellanca	Super Viking	300
Cessna	185	300
Cessna	Centurion	300
Cessna	Stationair 6	300
Cessna	Stationair 8	300
Piper	Saratoga	300
Piper	Turbo Saratoga	300
Piper	Saratoga SP	300
Piper	Turbo Saratoga SP	300
Cessna	Pressurized Centurion	310
Cessna	Turbo Centurion	310
Cessna	Turbo Stationair 6	310
Cessna	Turbo Stationair 8	310

Cessna 152—the lowest powered production aircraft—has 110 hp. (Cessna photo)

It takes 310 hp to propel the Cessna Turbo Stationair 8. (Cessna photo)

The Wing Derringer has two 160-hp Lycomings. (Wing photo)

■ TWIN-ENGINE AIRCRAFT

Manufacturer	Type	HP
Wing	Derringer	320
Beech	Duchess 76	360
Piper	Seminole	360
Piper	Seneca II	400
Cessna	Skymaster	420
Cessna	Turbo Skymaster	420
Cessna	Pressurized Skymaster	450
Piper	Aztec F	500
Piper	Turbo Aztec F	500
Beech	Baron B 55	520
Beech	Baron 58	570
Beech	Baron E 55	570
Cessna	310	570
Cessna	Turbo 310	570
Piper	Aerostar 600 A	580
Piper	Aerostar 601 B	580
Piper	Aerostar 601 P	580
Cessna	335	600
Cessna	340 A	620
Cessna	Chancellor	620
Piper	Navajo C	620
Beech	Baron 58 TC	650
Beech	Baron 58P	650
Cessna	402 C	650
Piper	Navajo C/R	650
Piper	Chieftain	700
Cessna	Golden Eagle	750
Cessna	Titan	750
Beech	Duke	760

The Beech Duke has two 380-hp turbocharged Lycomings. (Beech photo)

Current Production Aircraft Ranked by Engine TBO

The aircraft manufacturer provides a suggested time between overhauls (TBO) *for the powerplant. This is not a mandatory period, but it gives you an idea of how much time can elapse before the manufacturer thinks the engine should be torn down for an overhaul. Engines that receive a lot of abuse may very well not make it to the TBO. Engines that are babied may well exceed it by a substantial margin.*

The Bellanca Aries T-250 has a 250-hp Lycoming, with a 2,000-hour TBO. (Bellanca photo)

■ SINGLE-ENGINE AIRCRAFT

Manufacturer	Type	TBO Hours
Beech	Skipper	2,000
Beech	Sundowner 180	2,000
Bellanca	Aries T-250	2,000
Bellanca	Citabria 7ECA	2,000
Bellanca	Decathlon	2,000
Bellanca	Scout	2,000
Cessna	152	2,000
Cessna	180	2,000
Cessna	Cutlass RG	2,000
Cessna	Hawk XP	2,000
Cessna	Skyhawk	2,000
Cessna	Skylane RG	2,000
Maule	Lunar Rocket	2,000
Piper	Archer II	2,000
Piper	Dakota	2,000
Piper	Saratoga	2,000

The Cessna Pressurized Centurion gets only 1,400 hours TBO on its Continental engine. (Cessna photo)

Manufacturer	Type	TBO Hours
Piper	Saratoga SP	2,000
Piper	Super Cub	2,000
Piper	Tomahawk	2,000
Piper	Warrior II	2,000
Socata	Rallye 235 GT	2,000
Varga	Kachina	2,000
Cessna	Turbo Skylane RG	1,800
Mooney	Turbo 231	1,800
Piper	Turbo Arrow IV	1,800
Piper	Turbo Dakota	1,800
Piper	Turbo Saratoga	1,800
Piper	Turbo Saratoga SP	1,800
Beech	Sierra 200	1,600
Lake	Buccaneer	1,600
Mooney	201	1,600
Piper	Arrow IV	1,600
Beech	Bonanza A 36	1,500
Beech	Bonanza F 33A	1,500
Beech	Bonanza V 35B	1,500
Bellanca	Super Viking	1,500
Cessna	185	1,500
Cessna	Centurion	1,500
Cessna	Hawk XP	1,500
Cessna	Skylane	1,500
Cessna	Stationair 6	1,500

Manufacturer	Type	TBO Hours
Cessna	Stationair 8	1,500
Beech	Bonanza A 36 TC	1,400
Cessna	Pressurized Centurion	1,400
Cessna	Turbo Centurion	1,400
Cessna	Turbo Stationair 6	1,400
Cessna	Turbo Stationair 8	1,400

The Beech Duchess 76, one of five light twins with 2,000-hour TBO engines. (Beech photo)

■ **TWIN-ENGINE AIRCRAFT**

Manufacturer	Type	TBO Hours
Beech	Duchess 76	2,000
Piper	Aerostar 600 A	2,000
Piper	Aztec F	2,000
Piper	Seminole	2,000
Wing	Derringer	2,000
Piper	Aerostar 601 B	1,800
Piper	Aerostar 601 P	1,800
Piper	Navajo C	1,800
Piper	Seneca II	1,800
Piper	Turbo Aztec F	1,800
Beech	Duke	1,600
Cessna	Golden Eagle	1,600
Piper	Chieftain	1,600
Piper	Navajo C/R	1,600
Beech	Baron 58	1,500
Beech	Baron B 55	1,500
Beech	Baron E 55	1,500
Cessna	310	1,500
Cessna	Skymaster	1,500
Beech	Baron 58 TC	1,400
Beech	Baron 58P	1,400
Cessna	335	1,400
Cessna	340	1,400
Cessna	402 C	1,400
Cessna	Chancellor	1,400
Cessna	Pressurized Skymaster	1,400
Cessna	Turbo 310	1,400
Cessna	Turbo Skymaster	1,400
Cessna	Titan	1,200

The Cessna Titan has twin Continentals, each with a 1,200-hour TBO. (Cessna photo)

Current Production Aircraft Ranked by 65% Power Cruise Speed at Optimum Altitude

*Optimum altitude is 8,000 feet for non-turbocharged aircraft and 18,000 feet for turbocharged aircraft, indicated with an asterisk**

■ SINGLE-ENGINE AIRCRAFT

Manufacturer	Type	Speed kts
Beech	Bonanza A 36 TC	175*
Bellanca	Aries T-250	174
Cessna	Pressurized Centurion	173*
Cessna	Turbo Centurion	173*
Mooney	Turbo 231	170*
Cessna	Centurion	164
Piper	Turbo Saratoga SP	164*
Beech	Bonanza F 33A	163
Beech	Bonanza V 35B	163
Bellanca	Super Viking	162
Piper	Turbo Arrow IV	161*
Cessna	Turbo Skylane RG	159*
Beech	Bonanza A 36	158
Mooney	201	157
Piper	Saratoga SP	149
Cessna	Skylane RG	148
Cessna	Turbo Stationair 6	148*
Piper	Turbo Dakota	146*
Cessna	Turbo Stationair 8	143*
Cessna	Stationair 6	140
Cessna	185	138
Piper	Saratoga	138
Piper	Dakota	137
Cessna	Stationair 8	136
Cessna	Skylane	135
Maule	Lunar Rocket	134
Cessna	180	133
Lake	Buccaneer	130
Piper	Arrow IV	128
Piper	Archer II	126
Beech	Sierra 200	125
Socata	Rallye 235 GT	125
Cessna	Hawk XP	119
Bellanca	Decathlon	115
Cessna	Skyhawk	115
Piper	Warrior II	112
Beech	Sundowner 180	108
Varga	Kachina	105
Bellanca	Citabria 7ECA	101
Bellanca	Scout	100
Cessna	152	100
Piper	Super Cub	100
Beech	Skipper	96
Piper	Tomahawk	92

The Beech Bonanza A36TC—the fastest single—cruises at 175 kts at 18,000 feet. (Beech photo)

The Piper Tomahawk takes you around the pattern at 92 kts. (Piper photo)

■ TWIN-ENGINE AIRCRAFT

Manufacturer	Type	Speed kts
Piper	Aerostar 601 B	220*
Piper	Aerostar 601 P	220*
Beech	Baron 58 TC	212*
Beech	Baron 58P	212*
Cessna	Golden Eagle	208*
Beech	Duke	207*
Piper	Aerostar 600 A	207
Piper	Chieftain	197*
Cessna	335	196*
Cessna	340 A	196*
Cessna	402 C	196*
Piper	Navajo C/R	196*
Cessna	Turbo 310	194*
Piper	Navajo C	192*
Cessna	Pressurized Skymaster	191*
Beech	Baron E 55	190
Beech	Baron 58	190
Piper	Turbo Aztec F	194*
Cessna	Chancellor	190*
Cessna	310	182
Wing	Derringer	182
Beech	Baron B 55	180
Cessna	Turbo Skymaster	180*
Piper	Seneca II	176*
Piper	Aztec F	169
Cessna	Skymaster	160
Beech	Duchess 76	156
Piper	Seminole	155

The Piper Aerostar 601P—the fastest piston twin, at 220 kts. (Piper photo)

The Piper Seminole—the slowest light twin, at 155 kts. (Piper photo)

Current Production Aircraft Ranked by Sea-Level Rate-of-Climb

These rate-of-climb figures are for performance in standard air at sea level. Unless the aircraft is turbocharged (indicated with an asterisk), the climb-rate will deteriorate rapidly with altitude.*

■ SINGLE-ENGINE AIRCRAFT

Manufacturer	Type	Rate of climb fpm
Maule	Lunar Rocket	1,350
Bellanca	Aries T-250	1,240
Lake	Buccaneer	1,200
Bellanca	Super Viking	1,170
Beech	Bonanza F 33A	1,167
Beech	Bonanza V 35B	1,167
Cessna	Skylane RG	1,140
Beech	Bonanza A 36 TC	1,122*
Piper	Turbo Saratoga SP	1,120*
Piper	Dakota	1,110
Cessna	180	1,100
Bellanca	Scout	1,090
Mooney	Turbo 231	1,080*
Cessna	185	1,075
Piper	Turbo Saratoga	1,075*
Cessna	Turbo Skylane RG	1,040*
Beech	Bonanza A 36	1,030
Mooney	201	1,030
Bellanca	Decathlon	1,025
Cessna	Skylane	1,010
Cessna	Turbo Stationair 6	1,010*
Piper	Saratoga	1,010
Piper	Saratoga SP	1,000
Socata	Rallye 235 GT	970
Piper	Super Cub	960
Cessna	Centurion	950
Piper	Turbo Arrow IV	940*
Cessna	Pressurized Centurion	930*
Cessna	Turbo Centurion	930*
Beech	Sierra 200	927
Cessna	Stationair 6	920
Varga	Kachina	910
Piper	Turbo Dakota	902*
Cessna	Turbo Stationair 8	885*
Cessna	Hawk XP	870
Piper	Arrow IV	831
Cessna	Stationair 8	810
Cessna	Cutlass RG	800
Beech	Sundowner 180	792
Cessna	Skyhawk	770
Piper	Archer II	735
Bellanca	Citabria 7ECA	725
Beech	Skipper	720
Piper	Tomahawk	718
Cessna	152	715
Piper	Warrior II	710

The Maule Lunar Rocket departs its hangar. (Maule photo)

The Piper Warrior has the lowest rate of climb of any production single—710 fpm. (Piper photo)

The Cessna Golden Eagle climbs at 1,940 fpm fully loaded. (Cessna photo)

■ TWIN-ENGINE AIRCRAFT

Manufacturer	Type	Rate of climb fpm
Cessna	Golden Eagle	1,940*
Piper	Aerostar 600 A	1,800
Cessna	Turbo 310	1,700*
Wing	Derringer	1,700
Beech	Baron B 55	1,693
Beech	Baron E 55	1,682
Cessna	310	1,662
Bech	Baron 58	1,660
Cessna	340 A	1,650*
Beech	Duke	1,601*
Cessna	Titan	1,575*
Beech	Baron 58P	1,529*
Cessna	Chancellor	1,520*
Piper	Turbo Aztec F	1,470*
Beech	Baron 58 TC	1,461*
Piper	Aerostar 601 B	1,460*
Piper	Aerostar 601 P	1,460*
Cessna	402 C	1,450*
Cessna	335	1,400*
Piper	Aztec F	1,400
Piper	Seminole	1,340
Piper	Seneca II	1,340*
Beech	Duchess	1,235
Piper	Navajo C	1,220*
Piper	Navajo C/R	1,200*
Cessna	Pressurized Skymaster	1,170*
Cessna	Turbo Skymaster	1,160*
Piper	Chieftain	1,120*
Cessna	Skymaster	940

The Cessna Skymaster climbs at 940 fpm. (Cessna photo)

Current Twin Engine Production Aircraft Ranked by Sea-Level Single-Engine Rate-of-Climb

The single-engine rate-of-climb performance of the general-aviation twins is not spectacular. These figures are for sea level, and unless the aircraft is turbocharged (indicated with an asterisk), the single-engine climb-rate will deteriorate rapidly with altitude.*

Manufacturer	Type	S/E Rate of climb fpm
Wing	Derringer	420
Beech	Baron B 55	397
Cessna	Turbo 310	390*
Beech	Baron E 55	388
Beech	Baron 58	380
Cessna	Pressurized Skymaster	375*
Cessna	310	370
Piper	Aerostar 600 A	360
Cessna	Golden Eagle	350*
Cessna	Turbo Skymaster	335*
Cessna	340	315*
Beech	Duke	307*
Cessna	402 C	301*
Cessna	Skymaster	300
Cessna	Chancellor	290*
Beech	Baron 58 TC	270*
Beech	Baron 58P	270*
Piper	Navajo C/R	255*
Piper	Navajo C	245*
Piper	Aerostar 601 B	240*

The Wing Derringer will climb 420 fpm with one feathered. (Wing photo)

The Cessna 335 will climb at only 200 fpm on one engine. (Cessna photo)

Manufacturer	Type	S/E Rate of climb fpm
Piper	Aerostar 601 P	240*
Piper	Aztec F	235
Cessna	Titan	230*
Piper	Chieftain	230*
Piper	Seneca II	225*
Piper	Turbo Aztec F	225*
Piper	Seminole	217
Beech	Duchess 76	201
Cessna	335	200

Inside the Cessna 335. (Cessna photo)

Current Production Aircraft Ranked by Endurance at 65% Power

This endurance figure is calculated from the total usable fuel, including any optional fuel, at the speed delivered by 65 percent power at optimum altitude. It includes an allowance for takeoff and climb.

■ **SINGLE-ENGINE AIRCRAFT**

Manufacturer	Type	Endurance (hours)
Cessna	Skylane	7.7
Bellanca	Scout	7.3
Cessna	180	7.3
Piper	Arrow IV	7.2
Cessna	Skylane RG	7.1
Cessna	Cutlass RG	7.0
Cessna	Hawk XP	6.9
Cessna	152	6.8
Cessna	Turbo Skylane RG	6.6
Cessna	Skyhawk	6.5
Piper	Turbo Arrow IV	6.5
Piper	Saratoga	6.4
Beech	Sierra 200	6.3
Beech	Sundowner 180	6.3
Cessna	Centurion	6.3
Mooney	Turbo 231	6.3
Piper	Saratoga SP	6.2
Cessna	Stationair 6	6.2
Piper	Warrior II	6.1
Bellanca	Aries T-250	6.0
Piper	Archer II	6.0
Piper	Dakota	5.9
Cessna	185	5.9
Cessna	Turbo Centurion	5.8
Cessna	Turbo Stationair 6	5.8
Mooney	201	5.8
Cessna	Press. Centurion	5.7
Lake	Buccaneer	5.6
Piper	Turbo Saratoga SP	5.5
Piper	Super Cub	5.3
Bellanca	Decathlon	5.2
Piper	Tomahawk	5.2
Socata	Rallye 235 GT	5.2
Bellanca	Super Viking	5.1
Cessna	Stationair 8	5.1
Maule	Lunar Rocket	5.0
Beech	Bonanza A 36	4.8
Beech	Bonanza F 33A	4.8
Beech	Bonanza V 35B	4.8
Bellanca	Citabria 7ECA	4.7
Cessna	Turbo Stationair 8	4.7
Varga	Kachina	4.5
Beech	Bonanza A 36 TC	4.2
Beech	Skipper	4.1

The Cessna Skylane—its 7.7 hours endurance probably exceeds that of its pilot! (Cessna photo)

The Beech Skipper trainer. At 4.1 hours endurance, it should probably be refuelled after every second lesson. (Beech photo)

The Cessna Titan will stay up for 7.4 hours. (Cessna photo)

■ TWIN-ENGINE AIRCRAFT

Manufacturer	Type	Endurance (hours)
Cessna	Titan	7.4
Cessna	Skymaster	7.3
Piper	Aztec F	7.3
Beech	Baron 58	7.2
Cessna	Turbo 310	7.1
Cessna	340 A	6.9
Cessna	Chancellor	6.7
Cessna	Golden Eagle	6.7
Cessna	310	6.6
Cessna	Turbo Skymaster	6.5
Cessna	335	6.4
Piper	Seneca II	6.4
Piper	Navajo C	6.3
Cessna	Press. Skymaster	6.2
Piper	Seminole	6.2
Beech	Baron E 55	6.1
Beech	Duke	6.0
Cessna	402 C	6.0
Beech	Baron B 55	5.9
Piper	Aerostar 600 A	5.8
Piper	Navajo C/R	5.8
Piper	Turbo Aztec F	5.8
Piper	Aerostar 601 B	5.7
Piper	Aerostar 601 P	5.7
Piper	Chieftain	5.5
Wing	Derringer	5.5
Beech	Duchess 76	5.3
Beech	Baron 58 TC	5.0
Beech	Baron 58P	5.0

The Beech Baron 58P. Put 5.0 hours fuel on the flight plan. (Beech photo)

Current Production Aircraft Ranked by Gas Mileage at 65% Power

This mileage figure is calculated by taking the cruise speed in knots at 65 percent power at optimum altitude and dividing it by the fuel flow in US gph at 65 percent power. The result gives us nautical miles per gallon.

■ **SINGLE-ENGINE AIRCRAFT**

Manufacturer	Type	Efficiency (NMPG)
Cessna	152	19.4
Piper	Tomahawk	19.0
Bellanca	Citabria 7ECA	18.9
Mooney	Turbo 231	17.0
Mooney	201	16.8
Beech	Skipper	16.5
Piper	Archer II	16.4
Bellanca	Decathlon	15.0
Cessna	Skyhawk	15.0
Piper	Super Cub	15.0
Piper	Warrior II	14.9
Cessna	Cutlass RG	14.6
Bellanca	Aries T-250	14.5
Varga	Kachina	14.0
Beech	Sierra 200	13.9
Lake	Buccaneer	13.9
Piper	Arrow IV	13.2
Cessna	Hawk XP	13.0
Cessna	Turbo Skylane RG	12.9
Piper	Turbo Arrow IV	12.7
Piper	Turbo Dakota	12.7
Piper	Dakota	12.6
Cessna	Skylane RG	12.5
Beech	Bonanza A 36 TC	12.2
Beech	Bonanza F 33A	12.2
Beech	Bonanza V 35B	12.2
Cessna	Skylane	12.1
Cessna	Turbo Centurion	12.1

At 19.4 nmpg, the Cessna 152 gets better mileage than many cars. (Cessna photo)

Manufacturer	Type	Efficiency (NMPG)
Beech	Bonanza A 36	12.0
Beech	Sundowner 180	12.0
Bellanca	Scout	12.0
Cessna	Centurion	12.0
Cessna	180	11.9
Cessna	Press. Centurion	11.9
Bellanca	Super Viking	11.7
Maule	Lunar Rocket	11.2
Cessna	Turbo Stationair 6	10.3
Cessna	Stationair 6	10.2
Cessna	185	10.1
Cessna	Stationair 8	10.0
Cessna	Turbo Stationair 8	10.0
Piper	Saratoga	9.4
Piper	Saratoga SP	9.3
Piper	Turbo Saratoga SP	9.3
Socata	Rallye 235 GT	9.3
Piper	Turbo Saratoga	8.9

The Piper Turbo Saratoga gets only 8.9 nmpg, but carries up to seven people. (Piper photo)

The Wing Derringer gets 11.5 nmpg. (Wing photo)

■ TWIN-ENGINE AIRCRAFT

Manufacturer	Type	Efficiency (NMPG)
Wing	Derringer	11.5
Piper	Seminole	9.1
Piper	Seneca II	8.6
Beech	Duchess 76	8.4
Cessna	Press. Skymaster	8.2
Cessna	Turbo Skymaster	8.2
Cessna	Skymaster	8.1
Beech	Baron B 55	7.9
Piper	Aerostar 601 B	7.3
Piper	Aerostar 601 P	7.3
Beech	Baron 58	7.1
Beech	Baron E 55	7.1
Piper	Navajo C/R	6.8
Cessna	310	6.7
Piper	Aerostar 600 A	6.7
Piper	Navajo C	6.7
Beech	Baron 58P	6.6
Beech	Baron 58 TC	6.6
Cessna	Turbo 310	6.6
Piper	Turbo Aztec F	6.6
Cessna	340 A	6.5
Piper	Aztec F	6.5
Piper	Chieftain	6.4
Cessna	335	6.3
Cessna	Chancellor	6.3
Cessna	402 C	6.0
Beech	Duke	5.7
Cessna	Golden Eagle	5.4
Cessna	Titan	5.0

The Cessna Titan gets 5 nmpg but carries a big load. (Cessna photo)

Current Production Aircraft Ranked by Seat-Miles-per-Gallon

This figure is calculated by taking the optimum cruise-speed fuel mileage (nautical miles per gallon) and multiplying it by the number of seats available on the aircraft, including the crew.

At almost double the fuel efficiency of a Boeing 747, the Cessna Stationair 8 gets 80 seat-nmpg. (Cessna photo)

The Bellanca Scout gets only 24 seat-nmpg. (Bellanca photo)

■ SINGLE-ENGINE AIRCRAFT

Manufacturer	Type	SNMPG
Cessna	Stationair 8	80.0
Cessna	Turbo Stationair 8	80.0
Beech	Sierra 200	79.2
Beech	Bonanza A 36 TC	73.2
Cessna	Turbo Centurion	72.6
Bellanca	Aries T-250	72.5
Beech	Bonanza A 36	72.0
Cessna	Centurion	72.0
Cessna	180	71.4
Cessna	Pressurized Centurion	71.4
Mooney	Turbo 231	68.0
Mooney	201	67.2
Piper	Saratoga	65.8
Piper	Archer II	65.6
Piper	Saratoga SP	65.1
Piper	Turbo Saratoga SP	65.1
Piper	Turbo Saratoga	62.3
Cessna	Turbo Stationair 6	61.8
Cessna	Stationair 6	61.2
Beech	Bonanza F 33A	61.0
Beech	Bonanza V 35B	61.0
Cessna	185	60.6
Cessna	Skyhawk	60.0
Piper	Warrior II	59.6
Cessna	Cutlass RG	58.4
Lake	Buccaneer	55.6
Piper	Arrow IV	52.8
Cessna	Hawk XP	52.0
Cessna	Turbo Skylane RG	51.6
Piper	Turbo Arrow IV	50.8
Piper	Turbo Dakota	50.8
Piper	Dakota	50.4
Cessna	Skylane RG	50.0
Cessna	Skylane	48.4
Beech	Sundowner 180	48.0
Bellanca	Super Viking	46.8
Maule	Lunar Rocket	44.8
Cessna	152	38.8
Piper	Tomahawk	38.0
Bellanca	Citabria 7ECA	37.8
Socata	Rallye 235 GT	37.2
Beech	Skipper	33.0
Bellanca	Decathlon	30.0
Piper	Super Cub	30.0
Varga	Kachina	28.0
Bellanca	Scout	24.0

■ TWIN-ENGINE AIRCRAFT

Manufacturer	Type	SNMPG
Piper	Chieftain	64.0
Piper	Seneca II	60.2
Piper	Navajo C/R	54.4
Piper	Navajo C	53.6
Cessna	Chancellor	50.4
Cessna	Titan	50.0
Cessna	Pressurized Skymaster	49.2
Cessna	Turbo Skymaster	49.2
Cessna	Skymaster	48.6
Cessna	402 C	48.0
Beech	Baron B 55	47.4
Piper	Aerostar 601 B	43.8
Piper	Aerostar 601 P	43.8
Cessna	Golden Eagle	43.2
Beech	Baron 58	42.6
Beech	Baron E 55	42.6
Cessna	310	40.2
Piper	Aerostar 600 A	40.2
Beech	Baron 58P	39.6
Beech	Baron 58 TC	39.6
Cessna	Turbo 310	39.6
Piper	Turbo Aztec F	39.6
Cessna	340 A	39.0
Piper	Aztec F	39.0
Cessna	335	37.8
Piper	Seminole	36.4
Beech	Duke	34.2
Beech	Duchess 76	33.6
Wing	Derringer	23.0

The Piper Navajo Chieftain gets 64 seat-nmpg, making it the twin mileage leader. (Piper photo)

The Wing Derringer gets only 23 seat miles per gallon. (Wing photo)

As a matter of more than passing interest, here are the average seat-miles-per-gallon (exluding crew) for U.S.-airline-operated wide-body aircraft during the first quarter of 1979:

Airbus Industrie A 300 B	45.4
Boeing 747	43.4
Lockheed L 1011	42.8
McDonnell-Douglas DC 10	41.9

Turboprops

A turboprop engine is a jet engine connected to a propeller. The main forward thrust is provided by the propeller, and only a small amount of thrust comes from the jet exhaust. Because turboprops, also called prop-jets, are turbine engines, rather than reciprocating or piston engines, they are very efficient and simple to operate. However, they burn a lot of fuel at low altitudes, and they cost a great deal more than a piston engine of equivalent power. This is the main reason you don't see any smaller aircraft with turboprops, which is a pity.

Wherever the expense can be justified, or is no object, as with government operations, you'll find turbine engines selected over piston engines—most modern helicopters are turbine powered, for example. Years ago the Ontario (Canada) Department of Lands and Forests re-equipped its entire fleet of De Havilland Beavers with turboprop engines. The U.S. Navy is now going all turbine, using the Beech T-34C single-engine turboprop trainer to start pilots off, and disposing of their old piston-engine T-34s and T-28s. So turboprops are to be found mostly in commercial and government operations and in the larger business aircraft.

One of a string of King Airs—the A 100. (Beech photo)

Beech Aircraft Corporation, Wichita, KS 67201, telephone (316) 681-7111, makes the top selling turboprop King Air series. This comes in several models, with either Pratt and Whitney or AirResearch engines. Prices range from about $700,000 for the C 90 to about $1,275,000 for the 200 model.

The lowest cost twin turboprops are built by **Piper Aircraft Corporation, Lock Haven, PA 17745, telephone (717) 748-6711.** These are the Cheyennes, which come

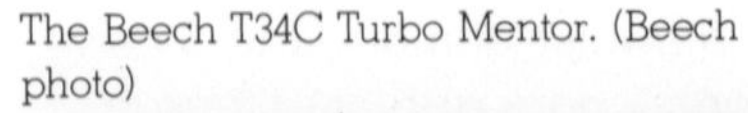

The Beech T34C Turbo Mentor. (Beech photo)

The top of the line from Piper—the Cheyenne III. (Piper photo)

in three models, all powered by Pratt and Whitney turbines. They range in price from $610,000 for the model I to about $1,100,000 for the Cheyenne III.

A unique propjet is the Hustler 500, built by **Gulfstream American Corporation, Box 2206, Savannah, GA 31402, telephone (912) 964-3000.** This has a turboprop in the nose and a jet in the tail. It has yet to achieve certification.

Rockwell International, General Aviation Division, 5001 North Rockwell Avenue, Bethany, OK 73008, telephone (405) 789-5000, introduced its new models 840 and 980 turboprops in late 1979. Both are powered by AiResearch engines.

Cessna Aircraft Company, Wichita, KS 67201, telephone (316) 685-9111, builds the Conquest, powered by AiResearch engines. They have now introduced a turboprop version of the 421 Golden Eagle, called the Corsair. This is powered by Pratt and Whitney 450 shp engines.

Swearingen Aviation Corporation, Box 32486, San Antonio, TX 78284, telephone (512) 824-9421, builds the Merlin twins. The original Merlin started out as an extension of the Beech Queen Air. It consisted of Queen Air wings and gear, with a new fuselage, tail unit and engine package. However, all the Merlins built today have brand new wings. There are two basic models, the IIIB, priced at around $1,600,000 and the stretched IVC, costing about $1,700,000. Both are powered by AiResearch engines.

Mitsubishi Aircraft International, Inc., 12700 Park Central Drive, Box 57, Dallas, TX 75251, telephone (214) 387-5600, builds two models in Japan, and these are assembled in the United States. They are known as MU-2s (Moo Twos), but are now marketed as the Marquise and the Solitaire.

There are some agricultural aircraft equipped with turboprops. See page 95 for more information.

The Rockwell Jetprop Commander 840 panel. (Rockwell International photo)

Inside the Cessna Conquest. (Cessna photo)

Current Production Turboprop Aircraft Ranked by Gross Weight

Most aircraft are rated by ramp weight, which includes the fuel required for taxiing. Take-off weight will be slightly lower. These figures are either ramp weight or gross weight, whichever is higher.

Manufacturer	Type	Weight lbs
Cessna	Corsair	8,275
Piper	Cheyenne I	8,750
Piper	Cheyenne II	9,050
Beech	King Air C 90	9,705
Cessna	Conquest 441	9,925
Beech	King Air E 90	10,160
Rockwell	Commander 840	10,375
Rockwell	Commander 980	10,375
Mitsubishi	Solitaire MU-2B-40	10,520
Beech	King Air F 90	11,030
Piper	Cheyenne III	11,080
Beech	King Air A 100	11,568
Mitsubishi	Marquise MU-2B-60	11,625
Beech	King Air B 100	11,875
Beech	Super King Air 200	12,590
Swearingen	Merlin IIIB	12,600
Swearingen	Merlin IVA	12,600

The Merlin IVA has a 12,6000-lb. ramp weight. (Swearingen photo)

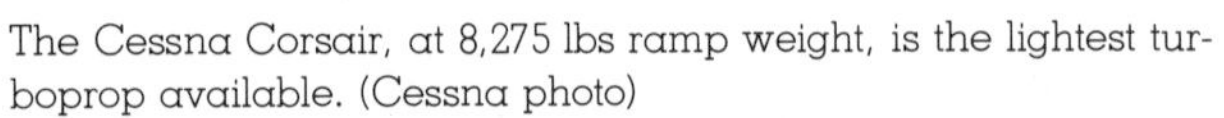
The Cessna Corsair, at 8,275 lbs ramp weight, is the lightest turboprop available. (Cessna photo)

Current Production Turboprop Aircraft Ranked by Useful Load

The figure shown is the maximum useful load available, excluding any optional equipment or fuel. In every case, the actual useful load of an aircraft will be lower than these figures indicate, reflecting radios and other options. However, it's all relative.

Manufacturer	Type	Useful Load lbs.
Beech	Super King Air 200	5,047
Swearingen	Merlin IIIB	4,800
Beech	King Air A 100	4,771
Beech	King Air B 100	4,763
Swearingen	Merlin IVA	4,400
Beech	King Air F 90	4,390
Rockwell	Commander 840	4,255
Cessna	Conquest 441	4,238
Beech	King Air E 90	4,108
Rockwell	Commander 980	4,104

The Beech Super King Air 200 has the best useful load—over 2½ tons. (Beech photo)

Manufacturer	Type	Useful Load lbs.
Piper	Cheyenne II	4,070
Mitsubishi	Marquise MU-2B-60	3,975
Beech	King Air C 90	3,927
Piper	Cheyenne I	3,846
Mitsubishi	Solitaire MU-2B-40	3,510
Cessna	Corsair	3,429

The Cessna Corsair light turboprop has a 3,429-lb. useful load. (Cessna photo)

Current Production Turboprop Aircraft Ranked by Seating Capacity

The figure shown is the total seating capacity, including crew. Turboprops are legal with one pilot, but are often flown with two. In any event, two of the seats counted are on the flight deck.

Manufacturer	Type	Seats
Beech	Super King Air 200	15
Swearingen	Merlin IVA	12
Cessna	Conquest 441	11
Mitsubishi	Marquise MU-2B-60	11
Rockwell	Commander 840	11
Rockwell	Commander 980	11
Swearingen	Merlin IIIB	11
Beech	King Air C 90	10
Beech	King Air E 90	10
Beech	King Air F 90	10
Beech	King Air A 100	10
Beech	King Air B 100	10
Mitsubishi	Solitaire MU-2B-40	9
Cessna	Corsair	8
Piper	Cheyenne II	8
Piper	Cheyenne III	8
Piper	Cheyenne I	7

The Beech Super King Air can carry 15 people. (Beech photo)

The Piper Cheyenne I can carry seven people. (Piper photo)

Current Production Turboprop Aircraft Ranked by Shaft Horsepower

The figure shown is the total shaft horsepower *available for takeoff. Since all the aircraft are twins, divide the figure by two to find the per-engine horsepower.*

Manufacturer	Type	SHP
Cessna	Corsair	900
Piper	Cheyenne I	1,000
Beech	King Air C 90	1,100
Beech	King Air E 90	1,100
Piper	Cheyenne II	1,240
Cessna	Conquest 441	1,272
Mitsubishi	Solitaire MU-2B-40	1,330
Beech	King Air A 100	1,360
Beech	King Air B 100	1,430
Mitsubishi	Marquise MU-2B-60	1,430
Rockwell	Commander 840	1,436
Piper	Cheyenne III	1,440
Rockwell	Commander 980	1,466
Beech	King Air F 90	1,500
Swearingen	Merlin IVA	1,680
Beech	Super King Air 200	1,700
Swearingen	Merlin IIIB	1,800

The Cessna Corsair is the lowest powered turboprop twin, at 900 shp. (Cessna photo)

The Swearingen Merlin IIIB has top power—1,800 shp. (Swearingen photo)

Current Production Turboprop Aircraft Ranked by Normal Cruising Speed at Optimum Altitude

Manufacturer	Type	Speed kts
Mitsubishi	Solitaire MU-2B-40	313
Rockwell	Commander 980	300
Mitsubishi	Marquise MU-2B-60	292
Swearingen	Merlin IIIB	287
Piper	Cheyenne III	283
Beech	Super King Air 200	279
Cessna	Conquest 441	271
Rockwell	Commander 840	268
Beech	King Air B 100	261
Beech	King Air F 90	258
Cessna	Corsair	257
Swearingen	Merlin IVA	247
Beech	King Air E 90	245
Piper	Cheyenne II	245
Beech	King Air A 100	242
Piper	Cheyenne I	234
Beech	King Air C 90	218

The Mitsubishi Solitaire cruises at 313 kts at 20,000 feet. (Mitsubishi photo)

A Beech King Air C 90 cruises at 218 kts at 18,000 feet. (Beech photo)

Current Production Turboprop Aircraft Ranked by Rate of Climb

The figure shown is the maximum sea level rate of climb at gross weight with both engines operating.

Manufacturer	Type	R/C
Rockwell	Commander 980	2,840
Rockwell	Commander 840	2,824
Piper	Cheyenne II	2,800
Swearingen	Merlin IIIB	2,780
Beech	Super King Air 200	2,450
Cessna	Conquest 441	2,435
Piper	Cheyenne III	2,400
Swearingen	Merlin IVA	2,400
Beech	King Air F 90	2,380
Mitsubishi	Solitaire MU-2B-40	2,350
Mitsubishi	Marquise MU-2B-60	2,200
Beech	King Air B 100	2,139
Beech	King Air A 100	1,963
Beech	King Air C 90	1,955
Cessna	Corsair	1,888
Beech	King Air E 90	1,870
Piper	Cheyenne I	1,750

The Rockwell Commander 980 climbs at 2,840 fpm. (Rockwell International photo)

The Piper Cheyenne I climbs at 1,750 fpm. (Piper photo)

Current Production Turboprop Aircraft Ranked by Single-Engine Rate of Climb

The figure shown is the maximum sea level single-engine rate of climb at gross weight.

Manufacturer	Type	S/E R/C
Rockwell	Commander 980	1,010
Rockwell	Commander 840	1,003
Beech	Super King Air 200	740
Swearingen	Merlin IIIB	723
Cessna	Conquest 441	715
Piper	Cheyenne II	660
Swearingen	Merlin IVA	650
Beech	King Air F 90	600
Piper	Cheyenne III	565
Beech	King Air C 90	539
Beech	King Air B 100	501
Mitsubishi	Solitaire MU-2B-40	475
Beech	King Air E 90	470
Beech	King Air A 100	452
Cessna	Corsair	424
Piper	Cheyenne I	413
Mitsubishi	Marquise MU-2B-60	410

The Marquise by Mitsubishi climbs at 410 fpm on one engine. (Mitsubishi photo)

The Rockwell Commander 980 climbs at over 1,000 fpm on one engine. (Rockwell International photo)

Current Production Turboprop Aircraft Ranked by Endurance at 65% Power

This endurance figure is calculated from the total usable fuel capacity, including any optional fuel, at the speed delivered by 65 percent power or equivalent at optimum altitude. It includes an allowance for takeoff and climb fuel.

Manufacturer	Type	Endurance (hours)
Cessna	Conquest 441	8.5
Swearingen	Merlin IIIB	8.4
Swearingen	Merlin IVA	8.0
Rockwell	Commander 840	7.3
Piper	Cheyenne II	6.6
Rockwell	Commander 980	6.1
Cessna	Corsair	5.7
Piper	Cheyenne I	5.5
Piper	Cheyenne III	5.5
Beech	King Air C 90	5.4
Beech	Super King Air 200	5.4
Beech	King Air E 90	5.3
Mitsubishi	Marquise MU-2B-60	5.3
Beech	King Air F 90	5.1
Beech	King Air B 100	5.0
Beech	King Air A 100	4.8
Mitsubishi	Solitaire MU-2B-40	4.4

The Cessna Conquest has up to 8.5 hours endurance. (Cessna photo)

The Mitsubishi Solitaire stays up 4.4 hours. (Mitsubishi photo)

Jets

The corporate jet aircraft is definitely here to stay. There are over 2,000 of them on the U.S. register, and U.S. manufacturers sold 233 new ones in 1978 for a total of about $620 million—or an average price of $2.7 million each. In fact *Business/Commercial Aviation* shows 1979 equipped prices ranging from almost $1.6 million for a Cessna Citation I to over $7.6 million for a Canadair Challenger. The *Aircraft Price Digest* lists the 1965 Gates Lear Jet 23 for about $565,000—about $30,000 less than its original price when new—and the 1965 Jet Commander 1121 for about $350,000.

Jets are expensive. No question. Their ownership is limited to corporations, charter companies, and superstars. I want one. And so does every pilot I know.

The Cessna Citation II, one of the best-selling bizjets. (Cessna photo)

The lowest-price corporate jet, as I mentioned above, is the Cessna Citation I. Cessna also builds the Citation II, and they are awaiting certification of the Citation III. The first two are the only business jets available today that legally may be flown by only one pilot (in which case, they are sub-designated "SP" models). This can be useful for certain operations, but two pilots—both qualified—are essential for most jet operations. Everything happens very fast in a jet, especially in an emergency situation.

The Cessna jets are a real success story, even though they fly substantially slower than their competition. Their rapid acceptance may be attributed to a combination of efficiency, excellent short-field performance, simplic-

The Cessna Citation III prototype. (Cessna photo)

The new Gates Learjet 55 (foreground) is much larger than the Model 35A. (Gates Learjet photo)

ity of design and handling, and, of course, the Cessna name. They are also very quiet, since they use fan jets, and are often allowed to land where other jets dare not tread. **Cessna Aircraft Company, Wichita, KS 67201, telephone (316) 685-9111.**

The name synonymous with bizjets is the Lear Jet, as invented by Bill Lear. Lear took a Swiss fighter plane, redesigned it, moved to Wichita, and built hundreds of them. Now the company is called **Gates Learjet Corporation, Box 7707, Wichita, KS 67277, telephone (316) 946-2000.** They make several varieties, including the new Longhorns, which incorporate NASA-type winglets.

The class act is the Gulfstream II, soon to be followed by the Gulfstream III. This is a large, spacious model that costs about $7 million. Production rights were sold by Grumman to American Jet Industries, and the resultant company is called **Gulfstream American, Box 2206, Savannah, GA 31402, telephone (912) 964-3000.**

Inside the Gates Learjet 35A. (Gates Learjet photo)

North American sold a lot of bizjets to the military and also made them available to the public as the Sabreliner. Now they are part of **Rockwell International Sabreliner Division, 6161 Aviation Road, St. Louis, MO 63134, telephone (314) 731-2260.**

Lockheed has for years offered the Jetstar—the only four-engine corporate jet. This originally started out as a twin but was produced only as a four-burner. It is no longer in production. **Lockheed-Georgia Company, 86 South Cobb Drive, Marietta, GA 30063, telephone (404) 424-9411.**

AiResearch Aviation offers the 731 Jetstar, which is basically a modified Lockheed Jetstar, with new Garrett TFE 731 fan jets and other improvements. **AiResearch Aviation Company, 6201 West Imperial Highway, Los Angeles, CA 90045, telephone (213) 646-5294.**

There once was an airplane called Jet Commander. When Aero Commander was acquired by Rockwell, builders of the Sabreliner, the U.S. Justice Department, in its infinite wisdom, decreed that either the Jet Commander or the Sabreliner had to be disposed of so that competition would be maintained in the industry. The JC was sold to Israel Aircraft Industries and, after some re-engineering, was renamed the Westwind, which is now marketed in the United States by **Atlantic Aviation Corporation, Box 15000, Wilmington, DE 19850, telephone (302) 322-7244.**

Several other aircraft built outside of the US have found a big following here. One is the Hawker Siddeley 125, marketed

The Dassault Falcon 10, smallest of the Falcons. (Falcon Jet photo)

The Falcon 50, almost a mini 727. (Falcon Jet photo)

by **British Aerospace, 2101 L Street NW, Washington, DC 20037, telephone (202) 857-0125.** And then there are the three Falcons, built by Avions Marcel Dassault and marketed here by **Falcon Jet Corporation, Teterboro, NJ 07608, telephone (201) 288-5300.** The smallest Falcon is the model 10. The model 20 is the original model, and the new 50 is a trijet very reminiscent of a mini Boeing 727.

Bill Lear designed another jet, which he called the LearLiner. This was the first "wide-bodied" bizjet. The aircraft was acquired by Canadair, it was changed about a bit, and it is now called the Challenger. **Canadair Ltd., Box 6087, Station A, Montreal, PQ H3C 3G9, Canada, telephone (514) 744-1511.** Their U.S. distributor is **Canadair, Inc., 274 Riverside Avenue, Westport, CT 06880, telephone (203) 226-1581.**

Mitsubishi Aircraft International, Inc. (12700 Park Central Drive, Box 57, Dallas, TX 75251, telephone (915) 387-5600) has announced the Diamond, a small fanjet, which will be built primarily in Japan and assembled in the U.S., in the same manner as their familiar MU-2 turboprops.

Finally, there is the Foxjet. This is a tiny, 4- to 6-place jet that, as of this writing, has not yet flown. If it goes, it will no doubt find a lot of appeal. However, the price has soared from the original $395,000 that was announced in 1977 to more than $800,000. **Foxjet International, Inc., 6701 West 110th Street, Minneapolis, MN 55438, telephone (612) 944-2255.**

The Mitsubishi Diamond will be on the market soon. (Mitsubishi photo)

This Foxjet mockup has been a popular attraction at air shows. (Foxjet photo)

Jet type ratings

You require a type rating on your pilot certificate before you can fly a jet as pilot in command. Unless you are a veteran and can get the training under the VA plan, to gain a jet type rating will cost you between $6,000 and $10,000, depending on your own ability, where you go, and what type of aircraft you use. Most pilots start out by getting a Cessna Citation or Lear Jet rating and obtaining a job flying copilot in one of these. To move into other aircraft, they usually join a firm that has both the aircraft they are rated on and something else. Then, after some experience is on the books on the rated aircraft, their employer will often provide the rating on the next aircraft at the employer's expense. In other words, it is difficult to get an employer to convert you to a jet rating if you're new, but once you're on the line, it's easier to move up at company expense. If your company is buying a new aircraft, the contract usually calls for the training of at least two pilots.

Here are some of the schools that offer jet type ratings, mostly on Citations and/or Lears:

American Flight Center
Building 2-S Meacham Field
Fort Worth, TX 76106
(817) 625-4149

Avia Corporation
1951 Airport Road
Atlanta, GA 30341
(404) 458-9921

Beckett-Tilford Inc.
Box 15887
West Palm Beach, FL 33406
(305) 683-4121
(800) 327-4349 (U.S., ex FL)

Burnside-Ott Aviation Training Center
Building 106, Opa Locka Airport
Miami, FL 33054
(305) 685-5111
(800) 327-5862 (U.S., ex FL)

Chipola Aviation, Inc.
Box 875
Marianna, FL 32446
(904) 482-8480

Danbury Airways, Inc.
Box 1254
Danbury, CT 06810
(203) 792-0100

Flight Proficiency, Inc.
Box 7510
Dallas, TX 75209
(214) 352-3901

FlightSafety International, Inc.
Marine Air Terminal
La Guardia Airport
New York, NY 11371
(212) 476-5700

Part of a jet rating is extensive ground school. This is at FlightSafety International. (FlightSafety International photo)

IASCO Aviation Training Division
100 Iasco Road
Napa, CA 94558
(415) 877-3600

Kal-Aero, Inc.
5605 Portage Road
Kalamazoo, MI 49002
(616) 343-2548

Jet Executive International
5601 NW 15th Avenue
Fort Lauderdale, FL 33309
(800) 327-5917 (U.S., ex FL)
(305) 776-4781 (FL)

Jet Fleet Corporation
Box 7445
Love Field, Dallas, TX 75209
(800) 527-6013 (U.S., ex TX)
(800) 492-4354 (TX)

Martin Aviation, Inc.
19331 Airport Way South
Santa Ana, CA 92707
(714) 546-4300
(800) 854-3666 (U.S., ex CA)

National Jets, Inc.
Box 22460
Fort Lauderdale, FL 33335
(305) 525-5538
(800) 327-3710 (U.S., ex FL)
(800) 432-2233 (FL)

National Jet Industries
19531 Airport Way South
Santa Ana, CA 92707
(714) 540-3930
(800) 854-6074 (U.S., ex CA)

Northern Air
Kent County Airport
Grand Rapids, MI 49508
(616) 949-5000

Sierra Academy of Aeronautics
Oakland International Airport
Oakland, CA 94614
(415) 568-6100

Western Skyways, Inc.
Portland-Troutdale Airport
Troutdale, OR 97060
(503) 665-1181

A book that may be helpful is *Your Jet Pilot Rating* by Frederick Bunyan (Modern Aircraft). Another is *Handling the Big Jets* by D. P. Davies (Pan American Navigation Service). The FAA requirements for type ratings are given in FAR 61.63.

Current Production Jet Aircraft Ranked by Basic Price (1980)

The price shown is the approximate basic 1980 price for the aircraft. Since most aircraft are ordered for delivery far into the future, the future price will be much higher.

Manufacturer	Type	Price
Cessna	Citation I	$1,470,000
Gates Learjet	24F	1,730,700
Gates Learjet	25D	1,797,500
Gates Learjet	Longhorn 28	1,970,000
Gates Learjet	Longhorn 29	2,025,600
Cessna	Citation II	2,045,000
Dassault	Falcon 10	2,300,000
Gates Learjet	35A	2,565,000
Israel Aircraft	Westwind I	2,639,000
Gates Learjet	36A	2,688,000
Israel Aircraft	Westwind II	3,147,500
British Aerospace	HS-125-700	3,540,000
Cessna	Citation III	3,795,000
Rockwell	Sabreliner 65	4,275,000
Dassault	Falcon 20F	5,800,000
Canadair	Challenger	6,000,000
Dassault	Falcon 50	7,442,000
Gulfstream American	Gulfstream III	9,750,000

Cessna Citation I is the lowest cost bizjet. (Cessna photo)

The Gulfstream III—almost $10 million. (Gulfstream-American photo)

Current Production Jet Aircraft Ranked by Gross Weight

Most aircraft are rated by ramp weight, which includes the fuel required for taxiing. Take-off weight will be slightly lower. These figures are either ramp weight or gross weight, whichever is higher.

Manufacturer	Type	Weight lbs
Cessna	Citation I	12,000
Cessna	Citation II	13,500
Gates Learjet	24F	13,800
Gates Learjet	25D	15,500
Gates Learjet	Longhorn 28	15,500
Gates Learjet	Longhorn 29	15,500
Gates Learjet	35A	17,250
Gates Learjet	36A	18,250
Dassault	Falcon 10	18,740
Cessna	Citation III	19,700
Israel Aircraft	Westwind I	23,000
Israel Aircraft	Westwind II	23,650
Rockwell	Sabreliner 65	24,000
British Aerospace	HS-125-700	25,000
Dassault	Falcon 20F	28,660
Canadair	Challenger	36,170
Dassault	Falcon 50	38,800
Gulfstream American	Gulfstream III	68,700

The Cessna Citation I is the lightest jet in production. (Cessna photo)

The Gulfstream III weighs in at over 34 tons. (Gulfstream-American photo)

Current Production Jet Aircraft Ranked by Useful Load

The figure shown is the maximum useful load available, excluding any optional equipment or fuel. In every case, the actual useful load of an aircraft will be lower than these figures indicate, reflecting radios and other options. However, it's all relative.

Manufacturer	Type	Useful Load lbs
Gulfstream American	Gulfstream III	36,200
Dassault	Falcon 50	18,960
Canadair	Challenger	18,650
Dassault	Falcon 20F	11,600
British Aerospace	HS-125-700	11,585
Israel Aircraft	Westwind II	10,810
Rockwell	Sabreliner 65	10,650
Israel Aircraft	Westwind I	10,610
Cessna	Citation III	9,715
Gates Learjet	36A	8,980
Gates Learjet	35A	7,979
Dassault	Falcon 10	7,940
Gates Learjet	25D	7,850
Gates Learjet	Longhorn 29	7,276
Gates Learjet	Longhorn 28	7,232
Gates Learjet	24F	6,736
Cessna	Citation II	6,319
Cessna	Citation I	5,443

The Gulfstream III carries over 18 tons of useful load. (Gulfstream-American photo)

The Cessna Citation I carries 5,443 lbs of useful load. (Cessna photo)

Current Production Jet Aircraft Ranked by Seating Capacity

The figure shown is the total possible seating capacity, including crew. *All jets, except certain Citations (the SP models), require two pilots. In the executive configuration there will be fewer seats.*

Manufacturer	Type	Seats
Canadair	Challenger	28
Gulfstream American	Gulfstream III	22
British Aerospace	HS-125-700	14
Dassault	Falcon 50	14
Dassault	Falcon 20F	12
Gates Learjet	35A	12
Israel Aircraft	Westwind I	12
Israel Aircraft	Westwind II	12
Rockwell International	Sabreliner 65	10
Gates Learjet	25D	10
Cessna	Citation II	10
Cessna	Citation III	10
Gates Learjet	Longhorn 28	10
Dassault	Falcon 10	10
Cessna	Citation I	8
Gates Learjet	24F	8
Gates Learjet	Longhorn 29	8
Gates Learjet	36A	8

You could pack 28 seats into a Canadair Challenger. (Canadair photo)

The Gates Learjet 36A carries eight. (Gates Learjet photo)

Current Production Jet Aircraft Ranked by Takeoff Thrust

The figure shown is the total thrust *available for takeoff. Since all the aircraft are twins, divide the figure by two to find the per-engine thrust.*

Manufacturer	Type	Thrust lbs
Cessna	Citation I	4,400
Cessna	Citation II	5,000
Gates Learjet	24F	5,900
Gates Learjet	25D	5,900
Gates Learjet	Longhorn 28	5,900
Gates Learjet	Longhorn 29	5,900
Dassault	Falcon 10	6,460
Gates Learjet	35A	7,000
Gates Learjet	36A	7,000
Cessna	Citation III	7,300
British Aerospace	HS-125-700	7,400
Israel Aircraft	Westwind I	7,400
Israel Aircraft	Westwind II	7,400
Rockwell	Sabreliner 65	7,400
Dassault	Falcon 20F	9,000
Dassault	Falcon 50	11,100
Canadair	Challenger	15,000
Gulfstream American	Gulfstream III	22,800

The Cessna Citation I—4,400 lbs of thrust. (Cessna photo)

The Gulfstream III punches out 22,800 lbs of thrust at takeoff. (Gulfstream-American photo)

Current Production Jet Aircraft Ranked by Normal Cruising Speed and Optimum Altitude

Manufactuer	Type	Speed kts
Canadair	Challenger	460
Dassault	Falcon 10	454
Dassault	Falcon 20F	453
British Aerospace	HS-125-700	445
Gulfstream American	Gulfstream III	445
Gates Learjet	24F	444
Gates Learjet	25D	444
Gates Learjet	Longhorn 28	441
Gates Learjet	Longhorn 29	441
Gates Learjet	35A	433
Gates Learjet	36A	433
Dassault	Falcon 50	432
Rockwell International	Sabreliner 65	430
Israel Aircraft	Westwind II	425
Israel Aircraft	Westwind I	424
Cessna	Citation III	420
Cessna	Citation II	385
Cessna	Citation I	354

The Canadair Challenger cruises at 460 kts. (Canadair photo)

The Cessna Citation I climbs at 2,680 fpm. (Cessna photo)

Current Production Jet Aircraft Ranked by Rate of Climb

The figure shown is the maximum sea level rate of climb at gross weight with both engines operating.

Manufacturer	Type	R/C
Gates Learjet	24F	7,100
Gates Learjet	25D	6,300
Gates Learjet	Longhorn 28	6,195
Gates Learjet	Longhorn 29	6,030
Canadair	Challenger	5,000
Gates Learjet	35A	4,900
Cessna	Citation III	4,815
Gates Learjet	36A	4,525
Dassault	Falcon 10	4,450
Israel Aircraft	Westwind I	4,000
British Aerospace	HS-125-700	3,900
Gulfstream American	Gulfstream III	3,800
Dassault	Falcon 20F	3,650
Israel Aircraft	Westwind II	3,600
Rockwell	Sabreliner 65	3,540
Dassault	Falcon 50	3,526
Cessna	Citation II	3,370
Cessna	Citation I	2,680

The Gates Learjet 24F can climb at over 7,000 fpm. (Gates Learjet photo)

The Cessna Citation I climbs 800 fpm on one engine. (Cessna photo)

Current Production Jet Aircraft Ranked by Single-Engine Rate of Climb

The figure shown is the maximum sea level single-engine rate of climb at gross weight.

Manufacturer	Type	S/E R/C
Gates Learjet	24F	2,050
Gates Learjet	Longhorn 28	1,900
Gates Learjet	Longhorn 29	1,900
Gates Learjet	25D	1,725
Canadair	Challenger	1,500
Gates Learjet	35A	1,500
British Aerospace	HS-125-700	1,350
Cessna	Citation III	1,340
Gates Learjet	36A	1,325
Gulfstream American	Gulfstream II	1,200
Cessna	Citation II	1,055
Dassault	Falcon 10	1,050
Rockwell	Sabreliner 65	950
Dassault	Falcon 20F	900
Israel Aircraft	Westwind I	860
Cessna	Citation I	800
Dassault	Falcon 50	NA
Israel Aircraft	Westwind II	NA

The Gates Learjet 24F can climb at over 2,000 fpm on one engine. (Gates Learjet photo)

The Cessna Citation I climbs 800 fpm on one engine. (Cessna photo)

Helicopters

Helicopters never made it into everyone's garage, as predicted in the 1930's, but they've certainly made it. Today's helicopter is a key factor in such varied enterprises as offshore-oil drilling, aerial photography, intracity transportation, business flying, construction, air ambulance, bank runs, logging, heavy-lift/external-cargo work, wildlife surveys, herding, coastal and border patrols, environmental protection, police work, mineral exploration, pipeline patrols, crop dusting, and much, much more.

Igor Sikorsky making his first helicopter flight in 1939. (Sikorsky photo)

Because of the Vietnam War, there are many helicopter pilots around. If you want to learn to fly one, be prepared to fork out a great deal of money—at least $90 an hour in the cheapest chopper, up to maybe $300 an hour in something like a Bell Jet Ranger, and you'll need at least 40 hours of instruction, with at least 15 of these being solo, to gain a certificate. For a commercial certificate, you need at least 150 hours of flight time as a pilot, with at least 50 of these being in helicopters.

Helicopters cost a ton of money to operate because they have many components that must be replaced at frequent intervals. The rotor blades and hub must be replaced every 2,000 hours or so in some models, for example. These are not cheap. And insurance rates are also very high. To charter a Jet Ranger might cost you around $300 to $400 an hour.

In the ten years from 1968 to 1978, annual sales of commercial helicopters in the United States soared from $57 million to $328 million. In 1978, U.S. manufacturers delivered 904 helicopters for commercial use, vs 522 in 1968. With the cessation of hostilities, military helicopter deliveries dropped from 2,800 to 166 in the same periods.

It's interesting to see how helicopters are regarded by various entities. When the PanAm building was built in New York, a key feature was the rooftop heliport, from which you could speed to nearby airports in a few minutes. I have landed and taken off on that pad many times and always have been enthralled by the concept. At one time helicopters were running from there to Teterboro, N.J. One day I had an 11 AM meeting in an office on Vanderbilt Avenue, a lunch date at Teterboro, and a 3 PM meeting on Madison Avenue. I left my 11 AM meeting, took the elevator to the top of the PanAm Building, choppered over to TEB, had lunch, and was back in Manhattan in good time for my next meeting. Talk about future shock! I was absolutely elated by the event. Lifting off from the PanAm Building is a mind-blowing experience, exceeded only by landing there.

A tragic accident a couple of years ago, though, closed down the PanAm heliport for good. But in Los Angeles, you are not al-

Bell's ubiquitous Jet Ranger. (Bell Helicopter photo)

lowed to put up a tall building unless you have a rooftop helipad, in case of emergency fire evacuations (remember the trouble they had in *The Towering Inferno*?). Washington, D.C., doesn't allow commercial helicopters to land downtown—only the President and his friends. And the government publishes a special helicopter VFR chart for the Los Angeles area.

I think it is impossible to look up for more than about ten minutes in New York, during daylight hours, without seeing at least one copter zipping by. The pads strategically located around Manhattan seem almost as precarious as a house in the Hollywood Hills—they are mostly situated half under bridges, abutments, or expressways. How the pilots maneuver their birds in and out escapes me.

The lowest-cost helicopter in production now is the new Robinson R 22, which costs about $44,000. It seats two and cruises at 92 mph. **Robinson Helicopter Co., Inc., 2247 Crenshaw Boulevard, Torrance, CA 90505, telephone (213) 539-0508.**

Another low-cost model is the Brantly-Hynes B2B, which runs around $60,000. It is also a two-seater. The same people also build a 5-seater, the 305, which costs about $100,000. **Brantly-Hynes Helicopter, Inc., Box 697, Frederick, OK 73542, telephone (405) 335-2256.**

The Brantly-Hynes B2B two-seater helicopter. (Brantly-Hynes photo)

The Bell 222 twin-turbine executive helicopter. (Bell Helicopter photo)

The Hughes 500D helicopter. (Hughes photo)

Other helicopter manufacturers are:

Bell Helicopter Textron
Box 482
Fort Worth, TX 76101
(817) 280-2011

Enstrom Helicopter Corp.
Box 277
Menominee, MI 49858
(906) 863-9971

Hiller Aviation
2075 West Scranton Avenue
Porterville, CA 93257
(209) 781-2261

Hughes Helicopters
Culver City, CA 90230
(213) 870-3361

Sikorsky Aircraft Division
Stratford, CT 06602
(203) 386-4000

There are a couple of choppers for home builders. One is the Scorpion 133, which is offered in kit form by **Rotorway, Inc., 14805 S Interstate 10, Tempe, AZ 85281, telephone (602) 963-6652.** It costs less than $15,000, but you have to build it yourself. It is powered by a liquid-cooled engine of their own design, which is now being offered to the homebuilder market. Another homebuilt is the Commuter IIA, offered by **International Helicopters, Inc., Box 107, Mayville, NY 14757, telephone (716) 753-2111.** This uses a Lycoming 150 hp engine.

Several foreign helicopters are to be found in world-wide service. The principal overseas manufacturers building helicopters for the civil market are:

The Sikorsky S 76 Spirit helicopter. (Sikorsky photo)

The Aerospatiale Dauphin 2 jet helicopter. (Aerospatiale photo)

Aerospatiale Helicopter Corporation
1701 West Marshall Drive
Grand Prairie, TX 75051
(214) 641-0000

Messerschmitt-Bolkow-Blohm GmBH
Box 514
Weston Way
West Chester, PA 19380
(215) 431-4150

Agusta S.p.A.
Atlantic Aviation Corporation
Box 1709
Wilmington, DE 19899
(302) 322-7000

Spitfire Helicopter Company, Box 61, Medina, PA 19063, telephone (215) 565-2986 is taking Polish-built, Russian-designed helicopters and upgrading and putting U.S. avionics in them. The result is offered as the Spitfire Taurus II twin-turbine helicopter. It has up to eight seats and sells for about $700,000.

The Helicopter Association of America, Suite 610, 1156 15th Street NW, Washington, DC 20005, telephone (202) 466-2420 represents the interests of the industry and its users. Every January the HAA has its annual convention and "industry showcase," usually in Las Vegas or some other western place of excitement and fantasy.

Rotor & Wing International is a monthly magazine catering to the chopper world. It is published by **PJS Publications Inc., Box 1790, Peoria, IL 61656, telephone (309) 682-6626.**

Current Production Helicopters Ranked by Basic Price (1980)

The price shown is the approximate basic price for the aircraft at the beginning of 1980.

Manufacturer	Type	Price
Robinson	R 22	43,795
Brantly-Hynes	B2B	57,450
Hughes	300C	86,000
Hiller	H 12E	98,000
Enstrom	Falcon F-28C-2	103,000
Brantly-Hynes	305	103,500
Enstrom	Shark 280C	108,000
Hiller	12 E4	109,000
Bell	Jet Ranger III	245,000
Hughes	500D	250,000
Aerospatiale	A-Star 350	277,000
Bell	Long Ranger II	390,000
Aerospatiale	Dauphin	675,000
MBB	BO-105 CBS	680,000
Agusta	109A	797,500
Bell	205A	895,000
Bell	222	975,000
Aerospatiale	Dauphin 2	1,100,000
Bell	212	1,140,000
Sikorsky	S 76 Spirit	1,275,000
Bell	214B	1,800,000

Current Production Helicopters Ranked by Gross Weight

Manufacturer	Type	Gross Weight lbs
Robinson	R 22	1,300
Brantly-Hynes	B2B	1,670
Hughes	300C	2,050
Enstrom	Falcon F-28C-2	2,350
Enstrom	Shark 280C	2,350
Hiller	12 E4	2,800
Brantly-Hynes	305	2,900
Hughes	500D	3,000
Hiller	H 12E	3,100
Bell	Jet Ranger III	3,200
Bell	Long Ranger II	4,050
Aerospatiale	A-Star 350	4,300
MBB	BO-105 CBS	5,070
Agusta	109A	5,730
Aerospatiale	Dauphin	6,615
Aerospatiale	Dauphin 2	7,495
Bell	222	7,650
Bell	205A	9,500
Sikorsky	S 76 Spirit	10,000
Bell	212	11,200
Bell	214B	13,800

Current Production Helicopters Ranked by Useful Load

This is the maximum useful load, including fuel and any optional equipment.

Manufacturer	Type	Useful Load lbs
Bell	214B	5,987
Bell	212	5,057
Bell	205A	4,308
Sikorsky	S 76 Spirit	3,650
Aerospatiale	Dauphin 2	3,042
Aerospatiale	Dauphin	2,665
Bell	222	2,500
Agusta	109A	2,090
Bell	Long Ranger II	1,890
MBB	BO-105 CBS	1,830
Aerospatiale	A-Star 350	1,800
Bell	Jet Ranger III	1,620
Hughes	500D	1,380
Hiller	H 12E	1,264
Brantly-Hynes	305	1,200
Hiller	12 E4	964
Hughes	300C	902
Enstrom	Falcon F-28C-2	758
Enstrom	Shark 280C	758
Brantly-Hynes	B2B	601
Robinson	R 22	500

Current Production Helicopters Ranked by Seating Capacity

Manufacturer	Type	Seats
Bell	214B	16
Bell	205A	15
Bell	212	15
Aerospatiale	Dauphin	14
Aerospatiale	Dauphin 2	14
Sikorsky	S 76 Spirit	14
Bell	222	10
Agusta	109A	8
Bell	Long Ranger II	7
Aerospatiale	A-Star 350	5
Bell	Jet Ranger III	5
Brantly-Hynes	305	5
Hughes	500D	5
MBB	BO-105 CBS	5
Hiller	12 E4	4
Enstrom	Falcon F-28C-2	3
Enstrom	Shark 280C	3
Hiller	H 12E	3
Hughes	300C	3
Brantly-Hynes	B2B	2
Robinson	R 22	2

Current Production Helicopters Ranked by Horsepower

In the case of twins (indicated with an asterisk), the horsepower shown is the total of both engines.*

Manufacturer	Type	HP
Robinson	R 22	124
Brantly-Hynes	B2B	180
Hughes	300C	190
Enstrom	Falcon F-28C-2	205
Enstrom	Shark 280C	205
Brantly-Hynes	305	305
Hiller	H 12E	305
Hiller	12 E4	305
Hughes	500D	420
Bell	Jet Ranger III	420
Bell	Long Ranger II	500
Aerospatiale	A-Star 350	615
Agusta	109A*	840
MBB	BO-105 CBS*	840
Aerospatiale	Dauphin	1,032
Bell	222*	1,230
Aerospatiale	Dauphin 2*	1,282
Sikorsky	S 76 Spirit*	1,300
Bell	205A	1,400
Bell	212*	2,580
Bell	214B	2,930

Current Production Helicopters Ranked by Cruising Speed at Optimum Altitude

Manufacturer	Type	Cruising Speed KTS
Bell	212	151
Sikorsky	S 76 Spirit	151
Aerospatiale	Dauphin	146
Bell	222	144
Hughes	500D	143
Agusta	109A	142
Bell	214B	140
Aerospatiale	Dauphin 2	136
Aerospatiale	A-Star 350	128
MBB	BO-105 CBS	126
Bell	Jet Ranger III	123
Bell	Long Ranger II	117
Bell	205A	112
Robinson	R 22	94
Brantly-Hynes	305	90
Enstrom	Falcon F-28C-2	88
Enstrom	Shark 280C	88
Hughes	300C	87
Brantly-Hynes	B2B	78
Hiller	H 12E	61
Hiller	12 E4	61

Current Production Helicopters Ranked by Sea-Level Rate of Climb

Manufacturer	Type	R/C
Bell	214B	2,200
Aerospatiale	Dauphin 2	2,010
Hughes	500D	1,900
Sikorsky	S 76 Spirit	1,800
Bell	222	1,730
Aerospatiale	A-Star 350	1,713
Bell	205A	1,680
Agusta	109A	1,620
MBB	BO-105 CBS	1,620
Bell	Long Ranger II	1,440
Bell	212	1,420
Aerospatiale	Dauphin	1,400
Brantly-Hynes	B2B	1,330
Bell	Jet Ranger III	1,260
Robinson	R 22	1,200
Enstrom	Falcon F-28C-2	1,100
Enstrom	Shark 280C	1,100
Hiller	H 12E	993
Hiller	12 E4	993
Brantly-Hynes	305	975
Hughes	300C	800

Current Production Helicopters Ranked by Endurance at Normal Cruise

Manufacturer	Type	Endurance hours
Bell	205A	5.1
Hughes	300C	4.3
Bell	212	4.1
Hiller	12 E4	3.9
Hiller	H 12E	3.9
Aerospatiale	A-Star 350	3.8
Sikorsky	S 76 Spirit	3.8
Aerospatiale	Dauphin	3.5
Bell	222	2.8
Bell	Jet Ranger III	2.7
Bell	Long Ranger II	2.7
Enstrom	Falcon F-28C-2	2.7
Enstrom	Shark 280C	2.7
Aerospatiale	Dauphin 2	2.6
Agusta	109A	2.6
MBB	BO-105 CBS	2.6
Brantly-Hynes	B2B	2.4
Robinson	R 22	2.4
Hughes	500D	2.1
Brantly-Hynes	305	1.9
Bell	214B	1.4

Gyrocopters

The gyrocopter has never taken off as a viable commercial enterprise, although this has been tried many times. Where are you now, Umbaugh, Cierva, Avian? Gyroplanes offer STOL performance without helicopter costs and complexity. Gyrocopters can't hover, fly backwards or sideways, or take off and land vertically. But they can come very close to this. A gyrocopter is a powered aircraft that uses an engine and propeller to provide forward motion and a free-wheeling rotor to provide lift. In some models, the rotor can be hooked up to the engine to provide a jump-start capability. The gyrocopter scene is dominated by Professor Igor Bensen, who sells plans and parts for a variety of homebuilt gyrocopters and gyrogliders.

The basic Bensen B-8M gyrocopter weighs 247 pounds empty and has a gross weight of 500 pounds. It has a rotor diameter of 260 inches, cruises at 60 mph, takes off at 20 mph, and lands at 7 mph. In calm air, it needs 300 feet to take off, and its landing roll is 20 feet. It has an endurance of 1.5 hours, will climb at 1,000 fpm, and has a service ceiling of 12,500 feet. It is powered by a McCulloch 4318E 4-cylinder, air-cooled, two-stroke engine. Construction plans cost about $35. **Bensen Aircraft Corp., Box 31047, Raleigh, NC 27612, telephone (919) 787-4224.**

Bensen supplies the parts; you put them together. (Bensen Aircraft photo)

Gyropilots are licensed by the FAA, just like regular airplane drivers. For a private pilot license with gyroplane rating, you must have a total of 40 hours instruction time in an aircraft, with at least 10 hours solo in a gyroplane, including flights with takeoffs and landings at paved and unpaved airports, 3 hours of cross-country flying, including a flight with landings at three or more points, each of which must have been 25 nautical miles from each of the other points. You must be at least 17 years old and pass the normal medical and written exam requirements for a private pilot certificate; there are some extra questions on rotorcraft in the exam. You also have to pass a flight test.

The Cessna Ag Truck. (Cessna photo)

Agricultural Aviation

The agricultural segment of aviation is so specialized that there are several aircraft designed exclusively for crop dusting and spraying. Ag flying takes special piloting skills. To fly a large, heavy aircraft low and slow, off dirt strips, for many hours a day with all the usual low-level hazards such as power lines, trees, towers, etc., makes life interesting, to say the least. Most U.S.-built agricultural aircraft are made by Cessna, Gulfstream American, Piper, and some specialty manufacturers.

Air Tractor, Inc. (Box 485, Olney, TX 76374, telephone (817) 564-5641) makes the Air Tractor 301 and 302 aircraft, powered by a 600-hp Pratt & Whitney radial and a 600-hp Lycoming turboprop, respectively, both with a 320-gallon hopper.

Cessna Aircraft Company (Box 1521, Wichita, KS 67201, telephone (316) 685-9111) offers three models, the AgWagon, powered by a 300-hp Continental, that can carry 200 gallons in its hopper; the AgTruck, a slightly larger version with the same power and a 280-gallon load capability; and the AgCarryall, a version of the 185 Skywagon, which doubles as a spray plane with a 151-gallon capacity, or a six-seater bushplane.

The Cessna Ag Carryall. (Cessna photo)

Ayres Corporation (Box 3090, Albany, GA 31706, telephone (912) 883-1440) took over the Thrush and the Turbo Thrush, the agplanes that used to be built by Aero Commander. The Thrush carries about 400 gallons of liquid and is powered by a 600-hp Pratt & Whitney radial engine. The Turbo Thrush is similar and is powered by a P & W PT-6 turboprop, and thus costs $100,000 extra.

The Ayres Turbo Thrush—a mean-looking machine. (Ayres photo)

Emair (Hangar 38, Harlingen Industrial Airpark, Harlingen, TX 78550, telephone (512) 425-6363) builds the Emair MA-1B, a biplane powered by a 1,200-hp Wright radial that will carry 475 gallons in its hopper.

Gulfstream American Corporation (Box 2206, Savannah, GA 31402, telephone (912) 964-3000) builds a variety of biplane agmachines, powered by 450-hp and 600-hp P & W radials and by the P & W PT-6 turboprop. These are called the Ag-Cats. The smaller ones carry 300 gallons, and the larger 500 gallons.

Piper Aircraft Corporation (Lock Haven, PA 17745, telephone (717) 748-6711) builds the Pawnee and Brave dusters. The

Piper's Pawnee D. (Piper photo)

Pawnee comes with a 235-hp Lycoming and a 150-gallon hopper, the Brave 300 has a 300-hp Lycoming and a 275-gallon hopper, while the Brave 375 has 75 more horsepower.

Weatherly Aviation Company, Inc. (2304 San Felipe Road, Hollister, CA 95023, telephone (408) 637-7354) sells the 201C, powered by a 450-hp P & W radial, with a 270-gallon-capacity hopper.

Agpilot training

Several operations specialize in training agpilots. Here are some of them:

Ag Pilot Training Center
Lehman Aviation, Inc.
Box 443
Casa Grande, AZ 85222
836-9388 in Arizona
(800) 528-3772 elsewhere.

Ayres Corporation Flight Training
Box 3090
Albany, GA 31706
(912) 883-1440

Benson Flying Service, Inc.
Municipal Airport
Benson, MN 56215
(612) 842-8441

Merigold Flying Service
Box 307
Merigold, MS 38759
(601) 748-2511

Books

Ag Pilot Flight Training Guide by David Frazier; *So You Think You Want to be a Crop Duster, Ag Aviation, All About Cropdusting, Agricultural Pilot and Chemicals, Ag Pilot Employment Guide*, all offered by **Diversified Publishing Company, 5301 44th Street, Lubbock, TX 79414;** *The Ag Pilot*, a magazine available at $10 a year from **Box 25, Milton-Freewater, OR 97682.**

Antique Airplanes

One of the best investments you can make today is in a Lockheed P 38 or a Beech Staggerwing. They'll never be built again, and there are only a few left in airworthy condition, so their value increases constantly. There are only a few Ford Trimotors flying now, and one of these was recently offered for sale for a million and a quarter bucks. Every issue of **Trade-A-Plane** (address **Crossville, TN 38555)** carries ads for the most unusual airplanes—here's a sampling from a recent issue of the antiques offered in the display ads:

Aircraft	Price offered
Beech D 17S Staggerwing—needs work	$25,000
Beech T-34A trainer	$40,000
Bucker Jungmann trainer	not priced
Cessna T-50 twin	$22,500
DH Vampire jet fighter	not priced
Douglas A-26 Invader bomber	$65,000
Fairchild F-22	$32,000
Fairchild F-24—1934	$14,500
Fairchild F-24R—1946	$19,900
Fairchild F-46A—1936	$19,950
Fairchild KR-34—1929—needs work	$10,500
Fieseler FI-156 Storch	$25,000
Ford Trimotor	$1.25 million
Grumman TBM torpedo bomber	$21,000
Lockheed T-33 jet trainer	not priced
Messerschmitt 208	not priced
Navy N3N biplane trainer	not priced
North American AT-6G trainer	$29,900

The Beech Staggerwing is a favorite with antiquers. (Beech photo)

North American B 25N bomber	not priced
North American T-6 trainer	$15,000
North American T-28C trainer	$69,500
North American SNJ-4 trainer	$21,000
Percival Proctor	not priced
Piper J-3 Cub—1943	$8,000
Republic P-47D Thunderbolt fighter	not priced
Republic Seabee amphibian	$57,000
Stinson 10A—1941—needs work	$1,800
Temco Swift	$33,000
Temco T-35 trainer	$19,500
Vickers Varsity transport	not priced
Waco—1941	$26,000

There are two associations for antique-airplane buffs: **Antique Airplane Association (Box H, Ottumwa, IA 52501, telephone (515) 938-2773)**, which publishes the *Antique Airplane Digest*, and the **Antique-Classic Division** of the Experimental Aircraft Association **(Box 229, Hales Corners, WI 53130, telephone (414) 425-4860**, which publishes *The Vintage Airplane*. You must be a member of EAA to join their Antique-Classic Division.

The Grumman F6F Hellcat—an antique-warbird lover's dream. (US Navy photo)

A Vickers Gun Bus at Rhinebeck puts on a lethal face. (Tim Foster photo)

Where to see antique airplanes

See the section on museums on page 202. If you live in the northeast, one of the best opportunities to see antique airplanes in action exists every weekend during the summer (May to October) at **Old Rhinebeck Aerodrome, Box 89, Rhinebeck, NY 12572, telephone (914) 758-8610.** This is located off Route 1, about 90 miles north of New York City. You wouldn't want to fly in to the aerodrome itself, since it's really out of the early days and isn't suitable for today's aircraft. However, you can fly in to **Skypark Airport** (46N) at Red Hook, NY, telephone **(914) 876-2880,** which is only a couple of miles north. They will run you over to Old Rhinebeck for a nominal charge. On Saturdays, Rhinebeck puts on its "Lindbergh Era" show, featuring aircraft from the 1920s and 30s, and on Sundays it's the World War I show, featuring, among others, a Fokker Triplane and a Sopwith Pup in a daring duel to the death. The shows run from 2:30 PM to 4:00 PM and there is a small admission charge. These shows are a lot of fun and well worth a visit.

Ground transportation between Old Rhinebeck and Skypark airports is rudimentary, but effective. (Tim Foster photo)

Cole Palen, founder and king of Old Rhinebeck Aerodrome. (Tim Foster photo)

Another location where you can see World War I airplanes in action is at the **Flying Circus Aerodrome** (3VA3), **Warrenton, VA, telephone (703) 439-8661.** Here on Sundays between mid-May and the end of October they put on a flying display starting at 2 PM. The airfield is all grass and is private. Landing there on Sundays requires prior permission. The airfield is on Route 17, 12 miles south of Warrenton.

Don't forget the EAA Convention at Oshkosh, WI, every August. Here you will see more old airplanes together in one place than anywhere else in the world. See page 177 for details.

The best information on where the old airplanes are is *Veteran & Vintage Aircraft,* compiled by Leslie Hunt, published by Scribners. This magnificent book literally takes you throughout the world, country by country, state by state, airfield by airfield, and tells you what old airplanes you can expect to see at each location. If you like old airplanes, you *must* get this book.

This Sopwith Camel may be seen at the Air Force Museum in Dayton, Ohio. (Air Force Museum photo)

This Cessna 195 would be any antiquer's delight. (Cessna photo)

Beautiful workmanship—the hallmark of many homebuilt airplanes. (Tim Foster photo)

Homebuilt Aircraft

There is an active homebuilt aircraft movement in North America. According to the **Experimental Aircraft Association (EAA),** there are now about 6,000 homebuilt aircraft flying in the United States, with another 1,200 in Canada. Thousands more are under construction in garages and lofts throughout the country. The EAA represents the interests of the homebuilders, as they are called. EAA's address is: **Box 229, Hales Corners, Wisconsin, 53130.** Their headquarters is located at **11311 West Forest Home Avenue, Franklin, Wisconsin, telephone (414) 425-4860.** Membership costs $25 per year, unless you are aged 18 or under, in which case it is $15 per year.

The EAA was started by Paul Poberezny in 1953, in Milwaukee, Wisconsin. It now has over 60,000 members and 600 local chapters. Its annual convention and fly-in, held at Oshkosh, Wisconsin, every August is the world's greatest air show for the true aviation enthusiast. About 25,000 members actually camp right at the airfield during the show. Throughout the week-long fly-in, over 10,000 aircraft visit and about 300,000 people attend. During that period, air traffic at Oshkosh is three times heavier than at Chicago's O'Hare!

The EAA includes three main divisions—the Antique/Classic, the Warbirds of America, and the International Aerobatic Club. Its 600 local chapters throughout the world promote sport aviation on the local level. Tied in with the chapters are 600 EAA Designee Inspectors, who serve as volunteers to assure high-quality standards in amateur-built aircraft construction, maintenance, and operation.

You can homebuild an aircraft from scratch to your own design, you can buy a set of plans for an existing design, or in some cases you can buy a kit of parts as well as the plans. However, the FAA has a rule that demands that "the majority" of an amateur-built aircraft must in fact be of the builder's own fabrication, so the kits do not permit the simple snapping together of a few interlocking parts.

There are about 200 different plans available, for everything from a single-seat ultra-light aircraft powered by two chainsaw motors to exotic scaled-down replicas of WW II fighters, such as the P-51 Mustang, the Hurricane, or the Focke-Wulf 190. Prices for the plans range from $10 to $500. The most popular aircraft, with over 450 now flying, is the Pitts Special—**Pitts Aviation Enterprises, Inc., Box 548, Homestead, FL 33030, telephone (305) 247-5423.**

Perhaps the most infamous kit-built aircraft is the Bede 5. This aircraft was launched by Jim Bede as a solution to the dream of a plane in every garage. The first flight of the prototype took place in 1971. Kits were offered for $2,000, and about 3,000 were sold. Many parts were mass-produced to handle the demand. However, not all the parts were built, and the engine was never made available.

In 1973 Bede announced that he would sell actual readymade Bede 5s—at $4,000 each! He asked for, and got, a deposit of $400 from anyone interested in a delivery position on the factory-built model. Over 6,000 people paid up—a total take of about $2.5 million.

The original engine was to be a 40-hp two-cylinder two-stroke engine. This proved unsatisfactory, so it was replaced in the design by the Japanese Xenoah, a three-cylinder, two-stroke, 70-hp motor. This engine was not yet FAA certified, so it could not be put into production aircraft.

The Bede 5 is certainly a sexy-looking machine—hardly a pilot worth his logbook wouldn't want to get his hands on it for a few minutes. However, Bede's imagination was substantially larger than his ability to deliver the real goods. He already had a reputation for "creative marketing"—making many promises that were seldom kept. Nevertheless, thousands of people paid for Bede 5 kits, and they are all still waiting for parts promised and paid for, but, as yet, not available. This has been going on for several years.

Meanwhile, Bede announced plans for a larger aircraft, the Bede 7. A few Bede 5s flew in the hands of Bede test pilots (one pilot was killed in an early flight), and a jet-powered version was developed. Five of these were built, and they were grouped up as an aerobatic team for the summer of 1975. One of the jets was bought by best-selling author Richard Bach, and recently one was for sale at an exclusive Beverly Hills shop called "The Toy Store"—at $50,000. Bede also built a Bede 5 and suspended it by a gantry from a truck to teach people its rather delicate handling techniques with greater safety. But no production Bede 5s were to be seen, and no complete kits were available. Only a handful of Bede 5s have flown; several of these have crashed. Ads in *Trade-a-Plane* offer partly built kits, delivery positions, and newsletters of hope and inspiration.

Legal wrangling has continued, with bankruptcies, consent decrees, liquidations, and other maneuvers taking place at frequent intervals. Many more BD-5s may actually fly, but it will only be thanks to the ardor of the frustrated builders, not to Jim Bede.

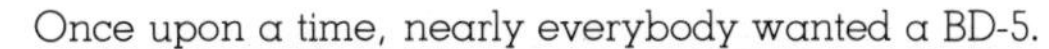

Once upon a time, nearly everybody wanted a BD-5.

The finest aircraft kit in the world—the Christen Eagle. (Christen Industries photo)

Another kit-built aircraft holds far more promise than the Bede. It is the Christen Eagle, a two-seater aerobatic biplane of great beauty. This has been hailed as a masterpiece by its buyers—everything fits, the instructions are extremely lucid and complete, and the whole project is well organized. The airplane was developed by Frank Christensen, who made millions from the semiconductor industry by the time he was in his early 30s. He started Christen Industries in 1971, and after trying unsuccessfully to buy the rights to manufacture the Pitts Special, he decided to design his own aerobatic biplane. It was rated 25 percent better than the Pitts in a comparison fly-off by aerobatic pilots Debbie Gary and Charlie Hillard in a *Flying* magazine article (April 1978). It would cost you about $25,000, plus your labor (about 1,400 hours), to buy and build your own Christen Eagle (1979 prices). **Christen Industries, Inc., 1048 Santa Ana Valley Road, Hollister, CA 95023, telephone (408) 637-7405.**

The superb Christen Eagle shows its paces. (Christen Industries photo)

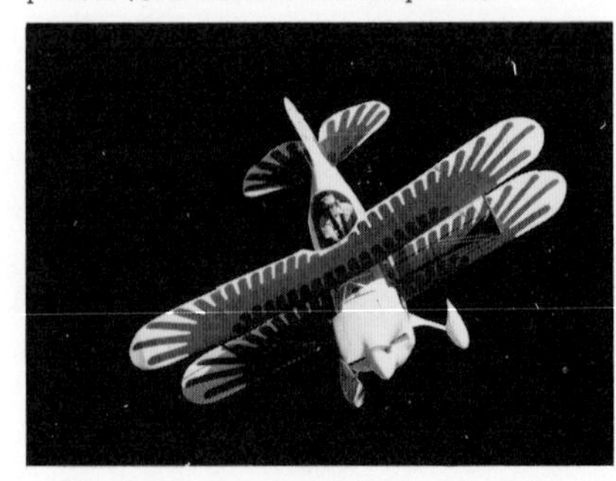

Rutan's VariViggen, his first canard design. (Rutan Aircraft photo)

The strangest-looking homebuilt aircraft are those designed by Burt Rutan. The most ubiquitous of these is the VariEze—somewhat reminiscent of the X-Wing fighter in *Star Wars.* This is a canard design—i.e., it flies tail-first. The fins and rudders are on the wing tips. You retract the nose gear to get in and out, and after take-off. The main gear stays down all the time. Electrical power can be drawn from solar cells on top of the fuselage. The engine is a 100-hp Continental 0-200, or a similar unit. A modified Volkswagen engine may be used. The aircraft, which is built of molded fiberglass, flies at 145 mph on less than 3 gallons per hour! It can be built in 600 hours, according to the kit seller, **Rutan Aircraft Factory, Building 13, Mojave, CA 93501, telephone (805) 824-2645.**

Another Rutan design is the "Quickie," a tiny aircraft powered by an 18-hp engine. This is a single-seater and is reported to be very easy to fly. It is also made from molded fiberglass and comes in partial kit form. Rutan claims it can be built in 400 manhours, and it gets 104 miles to the gallon! **Quickie Aircraft Construction, Box 786, Mojave, CA 93501, telephone (805) 824-4313.**

For people who hanker after a warbird, but can't face the prospect of huge fuel and maintenance costs, there is a variety of reduced-scale replicas available. **War Aircraft Replicas (348 South 8th Street, Santa Paula, CA 93060, telephone (805) 525-8212)** offers plans for the FockeWulfe 190, F4U Corsair, P 51 Mustang, P 47 Thunderbolt, Hawker Sea Fury, Curtiss P 40, Japanese Zero, plus others.

There is even an association of people who build and fly warbird replicas: **Replica Fighters of America (2789 Mohawk Lane, Rochester, Michigan 48063),** which costs $6 to join. And there's an outfit called **Air Replica International (Box 2218, Durango, CO 81301),** which is described as a specialized service organization for people who design or build replicas of famous aircraft. Their fee is $9 a year, which includes a subscription to their journal *Replicair.*

You mount a VariEze the same way you mount a camel. It kneels for you! (Tim Foster photo)

Rutan's weird and wonderful VariEze, a very popular homebuilt. (Rutan Aircraft photo)

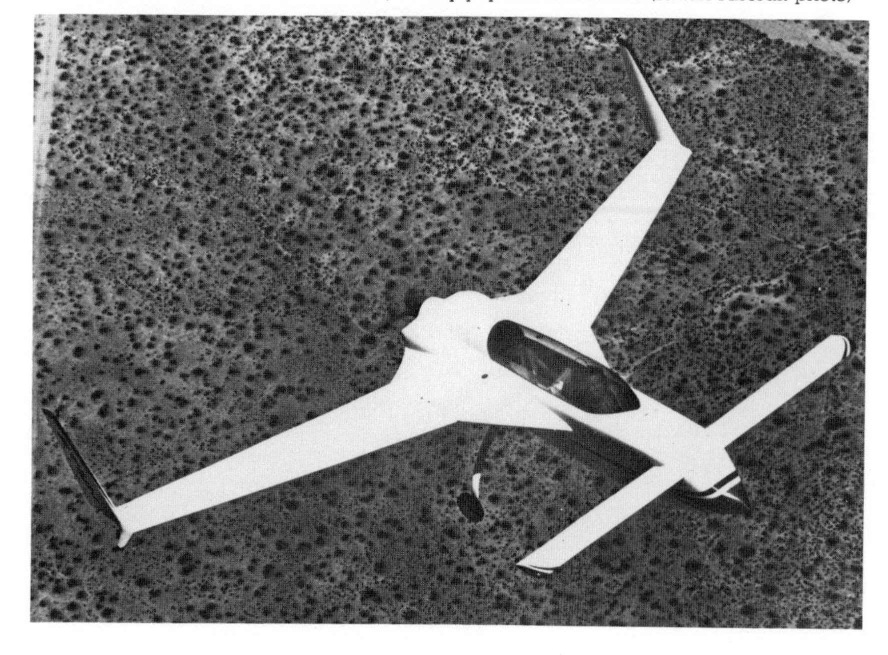

Molt Taylor (Box 1171, Longview, WA 98632, telephone (206) 425-9874) has designed a very sleek single-seater, the Mini-Imp, which looks like a sub-miniature Lear-Fan as it was originally conceived, complete with inverted V-tail. It can accommodate engines from 65 hp up to 115 hp. A two-place version is being studied.

Barney Oldfield Aircraft Co., Box 5974, Cleveland, OH 44101, telephone (216) 423-3816 supplies plans for the Baby Great Lakes biplane.

Not a Lear Fan, but a Taylor Mini-Imp homebuilt. (Rutan Aircraft photo)

A Barney Oldfield Baby Lakes homebuilt. (Barney Oldfield photo)

The EAA publishes a book *Sport Aircraft You Can Build* listing 172 designs of aircraft for which plans are available. This book is available from EAA at $4.95. EAA publishes many other books to assist the homebuilder. Here are some of the titles:

- *Basic Hand Tools, I and II*
- *Custom Aircraft Building Tips, I, II, III and IV*
- *Custom Built Sport Aircraft Handbook*
- *Design, I, II and III*
- *Engineering for the Custom Aircraft Builder*
- *Engine Operation, Carburetion, Conversion*
- *Engines, I and II*
- *Metal Aircraft Building Techniques*
- *Modern Aircraft Covering Techniques*
- *Pilot Reports and Flight Testing*
- *Sheet Metal, I and II*
- *Tips on Aircraft Fatigue*
- *Welding*
- *Wood, I and II*
- *Wood Aircraft Building Techniques*

These are all available from EAA at $3.65 each **(EAA, Box 229, Hales Corners, WI 53130).**

The Sportplane Builder, by **Tony Bingelis** is available from him at **8509 Greenflint Lane, Austin, TX 78759,** for $14.95 including postage.

Here is a list of FAA Advisory Circulars that should prove useful to the homebuilder:

■ **ACs THAT COST MONEY**

AC Number	Title
AC 20-9	Personal Aircraft Inspection Handbook (out of print—being revised)
AC 43.13-1A	Acceptable Methods, Techniques and Practices—Aircraft Inspection and Repair (price $3.70)
AC 43.13-1A Chg 1	Change 1 to the above (price $.65)
AC 43.13-1A Chg 1	Change 2 to the above (price $.35)
AC 43.13-2A	Acceptable Methods, Techniques and Practices—Aircraft Alterations (price $2.75)
AC 91.23A	Pilot's Weight and Balance Handbook (price $2.30)

■ **ACs THAT ARE FREE**

AC Number	Title
AC 00-2	Advisory Circular Checklist
AC 00-44	Status of Federal Aviation Regulations
AC 20-5D	Plane Sense
AC 20-27B	Certification and Operation of Amateur Built Aircraft
AC 20-28A	Nationally Advertised Construction Kits, Amateur-Built Aircraft
AC 20-32B	Carbon Monoxide (CO) Contamination in Aircraft—Detection and Prevention
AC 20-35B	Tie-Down Sense
AC 20-44	Glass Fiber Fabric for Aircraft Covering
AC 20-86	Aviation Education Through Building an Airplane
AC 21-12	Application for US Airworthiness Certificate FAA Form 8130-6
AC 21-13	Standard Airworthiness Certification of Surplus Military Aircraft and Aircraft Built From Spare and Surplus Parts
AC 43-5	Airworthiness Directives for General Aviation Aircraft
AC 43-9A	Maintenance Records: General Aviation Aircraft
AC 43-12	Preventive Maintenance
AC 43-16	General Aviation Airworthiness Alerts
AC 45-2	Identification and Registration Marking
AC 60-6A	Airplane Flight Manuals, Approved Manual Materials, Markings and Placards
AC 91-13B	Cold Weather Operation of Aircraft

To obtain Advisory Circulars that cost money, I suggest that first you get the latest copy of AC 00-2—*Advisory Circular Checklist*, which is free. This will give the latest price and availability of ACs, as well as current ordering information. Then follow the instructions given therein. If you can't obtain the *Checklist* at your local Flight Service Station, you can get it, and all the other free ACs mentioned, from:

Department of Transportation
Publications Section M-443.1
Washington, DC 20590.

Repairman's certificate

The FAA now allows a person who has built an aircraft successfully to apply for a *Repairman's Certificate*. This will be issued on application, using FAA Form 8310-2, provided that you can prove you built your airplane. The aircraft logbook should give enough evidence of this. Once you have this certificate, *which will be limited to the aircraft you built*, you are entitled to perform all maintenance on your aircraft, including the new annual *condition inspection* mentioned above. If you don't have the repairman's certificate, the condition inspection must be performed by a qualified A & P mechanic or FAA Inspector.

Used Aircraft

There's a man down in Oke City who runs a terrific operation. He keeps the industry informed about used aircraft prices and much more. Four times a year he issues the *Aircraft Price Digest*, which is the "bible" of the industry—employed by bankers, insurance companies, and aircraft dealers to help in gauging the fair price for used aircraft. He is **Bernie McGowan**, and his operation is located right on the grounds of the FAA Aeronautical Center at Will Rogers Field in Oklahoma City. He also offers a title search service and an aircraft appraisal service, and he organizes plane auctions. He is at Box 59977, Oklahoma City, OK 73159, telephone (405) 682-1471.

With Bernie McGowan's permission, I computerized the used-aircraft prices from the May 1979 *Aircraft Price Digest*, and organized the data by *price*. This was then sorted by year, then alphabetically by manufacturer and aircraft type. The lowest-priced aircraft come first, and for models with the same price, the most recent year comes first, and then it's alphabetical by maker. When you read this, it will be much later than May 1979; however, it's all relative, so it should still be quite helpful. Copyright © 1979, Aircraft Appraisal Association of America.

A Cessna 120—still a good used buy for about $5,000. (Cessna photo)

A 1973 Cessna 150, probably well-used as a trainer, makes an inexpensive first plane. (Cessna photo)

Used Aircraft Prices—May 1979

■ SINGLE ENGINE—TWO SEATS

Under $5,000

3,900	48	Luscombe 8 A
4,000	46	Aeronca Chief
4,250	48	Aeronca Champion
4,250	46	Ercoupe
4,500	60	Cessna 150
4,500	50	Aeronca Champion
4,500	48	Ercoupe
4,750	62	Cessna 150
4,750	62	Piper Colt
4,750	60	Forney Aircoupe

$5,000–$9,950

5,000	48	Cessna 120
5,000	46	Luscombe 8 E
5,100	66	Alon Aircoupe
5,250	64	Cessna 150
5,500	58	Bellanca Champion
5,500	48	Cessna 140
5,750	48	Luscombe 8 F
6,000	66	Cessna 150
6,000	60	Bellanca Champion
6,000	60	Luscombe S-86
6,000	50	Cessna 140
6,000	48	Piper PA 14 Cruiser
6,250	50	Piper Super Cub
6,500	68	Cessna 150
6,500	60	Bellanca Champion
6,750	70	Gulfstream American Clipper
6,750	62	Bellanca Champion
6,750	56	Piper Super Cub
6,850	70	Cessna 150
7,000	64	Bellanca Citabria
7,400	66	Bellanca Citabria
7,500	72	Cessna 150
7,750	72	Gulfstream American Trainer
7,750	68	Bellanca Citabria
8,000	74	Cessna 150
8,250	72	Gulfstream American Trainer
8,250	70	Bellanca Citabria
8,750	72	Bellanca Citabria
9,000	74	Gulfstream American Trainer
9,000	68	Beech Sport III
9,250	76	Cessna 150
9,500	70	Bellanca Citabria
9,750	74	Bellanca Citabria

$10,000–$14,950

10,000	74	Gulfstream American Trainer
10,000	74	Taylorcraft F-19
10,000	70	Beech Sport
10,000	68	Bellanca Citabria
10,000	50	Globe Swift
10,250	68	Bellanca Citabria
10,500	72	Bellanca Citabria
10,500	70	Bellanca Citabria
10,500	69	Bellanca Citabria
10,750	70	Bellanca Citabria
11,250	76	Taylorcraft F-19
11,250	58	Piper Super Cub
11,250	72	Bellanca Citabria
11,250	72	Beech Sport
11,500	76	Gulfstream American Trainer
11,500	74	Bellanca Citabria
11,500	72	Bellanca Citabria
11,500	62	Piper Super Cub
11,750	76	Bellanca Citabria
11,750	64	Piper Super Cub
12,000	76	Gulfstream American Trainer
12,000	66	Piper Super Cub
12,500	74	Bellanca Citabria
12,500	72	Bellanca Decathlon
12,750	74	Beech Sport
12,750	68	Piper Super Cub
13,000	74	Bellanca Citabria
13,250	78	Taylorcraft F-19
13,500	78	Cessna 152
13,500	70	Piper Super Cub
13,750	76	Bellanca Citabria
14,000	74	Bellanca Scout
14,000	72	Piper Super Cub

Price	Year	Aircraft
14,500	76	Bellanca Citabria
14,500	74	Bellanca Decathlon

$15,000–$19,950

Price	Year	Aircraft
15,000	76	Bellanca Citabria
15,500	78	Gulfstream American T-Cat
16,000	78	Piper Tomahawk
16,000	74	Piper Super Cub
16,250	78	Bellanca Citabria
17,000	76	Bellanca Scout
17,000	76	Beech Sport
17,250	76	Bellanca Decathlon
17,900	78	Bellanca Citabria
18,750	76	Piper Super Cub

Over $20,000

Price	Year	Aircraft
21,000	78	Varga Kachina
22,750	78	Piper Super Cub

■ SINGLE ENGINE—MORE THAN TWO SEATS—FIXED GEAR

Under $10,000

Price	Year	Aircraft
4,750	52	Piper Tri Pacer
5,500	54	Piper Tri Pacer
6,000	46	Stinson 108-1
6,500	56	Piper Tri Pacer
6,750	58	Piper Tri Pacer
7,000	56	Cessna 172
7,250	62	Maule M-4
7,250	48	Stinson 108-1
7,500	66	Aero Commander 100
7,500	58	Cessna 172
7,750	68	Aero Commander Darter
7,750	64	Maule M-4
8,000	68	Mooney Cadet
8,000	60	Cessna 172
8,000	58	Cessna 175
8,250	66	Maule M4-C
8,250	66	Beech Musketeer A 23 Sport
8,250	64	Piper Cherokee 140
8,500	48	Cessna 170
8,750	66	Piper Cherokee 140
8,750	64	Beech Musketeer
8,750	62	Piper Cherokee 160
8,750	60	Cessna 175 Skylark
9,000	70	Mooney Cadet
9,000	64	Piper Cherokee 160
9,000	50	Cessna 170
9,250	68	Piper Cherokee 140
9,250	66	Beech Musketeer
9,250	62	Cessna Skyhawk
9,500	66	Piper Cherokee 160

$10,000–$14,950

Price	Year	Aircraft
10,000	70	Piper Cherokee 140
10,000	66	Beech Musketeer
10,000	62	Cessna 175 Skylark
10,000	54	Cessna 170
10,250	68	Beech Custom
10,250	64	Cessna Skyhawk
10,450	70	Maule M-4-180C
10,700	60	Piper Tri Pacer
11,000	72	Piper Cherokee 140
11,000	66	Cessna Skyhawk
11,250	68	Maule M-4-210C
11,500	68	Aero Commander 180
11,500	64	Mooney Master
11,500	56	Cessna 182
12,000	70	Aero Commander 180
12,000	56	Cessna 170
12,250	68	Cessna Skyhawk
12,500	58	Cessna Skylane
13,000	74	Piper Cherokee Cruiser
13,000	64	Piper Cherokee 180
13,250	68	Cessna Cardinal
13,250	66	Beech Musketeer
13,500	70	Cessna Skyhawk
13,750	72	Gulfstream American Traveler
14,000	66	Piper Cherokee 180
14,500	72	Cessna Skyhawk
14,500	70	Beech Custom
14,500	68	Beech Musketeer
14,500	60	Cessna Skylane

$15,000–19,950

Price	Year	Aircraft
15,250	76	Piper Cherokee Cruiser
15,250	72	Maule M4-220C
15,250	64	Piper Cherokee 235
15,500	56	Cessna 180
15,750	74	Gulfstream American Traveler
15,750	72	Beech Sundowner C 23
15,750	62	Cessna Skylane
16,000	74	Cessna Skyhawk
16,000	68	Piper Cherokee 180
16,000	58	Cessna 180
16,250	74	Piper Cherokee Warrior
16,500	78	Gulfstream American Lynx
16,500	70	Cessna 177 Cardinal
16,500	66	Piper Cherokee 235
16,500	62	Cessna 185
16,500	60	Cessna 180
17,000	64	Cessna Skylane
17,000	62	Cessna 180
17,250	70	Piper Cherokee 180
17,500	74	Maule M-4-210C
17,750	68	Piper Cherokee 235
17,750	64	Cessna 185
18,000	66	Cessna Skylane
18,000	64	Cessna 180
18,250	76	Cessna Skyhawk
18,500	76	Gulfstream American Cheetah
18,500	76	Piper Cherokee Warrior
18,500	74	Beech Sundowner
18,750	72	Piper Cherokee 180
19,000	72	Cessna 177 B Cardinal
19,000	66	Cessna 180H
19,500	68	Cessna 182 L Skylane

$20,000–$29,950

Price	Year	Aircraft
20,000	68	Cessna 180 H
21,500	70	Cessna 180
21,500	70	Cessna Skylane
21,500	70	Piper Cherokee 235
21,500	64	Cessna 206
22,000	66	Piper Cherokee Six 260

Price	Year	Aircraft
22,750	68	Cessna 185
23,000	74	Piper Cherokee Archer
23,250	66	Piper Cherokee Six 300
23,750	72	Cessna 180
23,750	72	Cessna Skylane
23,750	66	Cessna 206
24,000	76	Beech Sundowner
24,500	78	Beech Sport
25,000	68	Cessna 206
25,000	66	Cessna 206 Turbo
25,500	76	Cessna 177 Cardinal
25,500	70	Cessna 185
25,750	76	Gulfstream American Tiger
26,000	78	Gulfstream American Cheetah
26,500	78	Cessna Skyhawk
26,750	74	Cessna Skylane
27,250	72	Cessna 185
27,500	68	Cessna 206 C Turbo
27,750	70	Cessna 206
28,000	74	Cessna 180
28,750	70	Piper Cherokee Six 300
29,000	74	Piper Cherokee Pathfinder
29,000	70	Cessna 207 Skywagon
29,750	78	Gulfstream American Tiger
29,750	72	Piper Cherokee Six 260

$30,000–$39,950

Price	Year	Aircraft
30,500	72	Cessna 206 Stationair
30,500	70	Cessna 206 Turbo
31,000	78	Cessna Hawk XP
31,000	74	Cessna Skywagon
31,500	70	Cessna 207 Turbo Skywagon
31,750	76	Cessna 180
31,750	72	Piper Cherokee Six 300
32,500	78	Piper Cherokee Archer II
32,750	76	Cessna Skylane
32,750	72	Cessna 207 Skywagon
33,000	72	Cessna 206 Turbo Stationair
33,500	78	Cessna Cardinal Classic
34,000	74	Piper Cherokee Six 260
34,250	76	Piper Cherokee Pathfinder
34,750	78	Beech Sundowner
35,000	74	Cessna 206 Stationair
35,000	74	Cessna 207 Skywagon
35,750	74	Piper Cherokee Six 300
36,000	76	Cessna Skywagon
36,000	72	Cessna 207 Turbo Skywagon
37,500	78	Cessna 180
37,500	74	Cessna 206 F Turbo Stationair

$40,000–$49,950

Price	Year	Aircraft
40,000	78	Cessna 182 Skylane
40,500	74	Cessna 207 Turbo Skywagon
41,000	78	Cessna Skywagon
41,500	76	Piper Cherokee Six 260
42,500	76	Cessna Stationair
43,500	76	Cessna 207 Skywagon
44,650	76	Cessna Turbo Stationair
47,000	76	Piper Cherokee Six 300

Over $50,000

Price	Year	Aircraft
52,500	78	Cessna Stationair 6
52,500	78	Cessna Stationair 7
52,500	78	Piper Cherokee Six 260
56,000	78	Cessna Turbo Stationair 6
57,500	78	Piper Cherokee Six 300
59,000	78	Cessna 207A Turbo Stationair

■ SINGLE ENGINE—RETRACTABLE GEAR

Under $10,000

Price	Year	Aircraft
8,000	46	Bellanca 14-13
9,000	56	Mooney 20
9,250	50	Bellanca 14-19

$10,000–$14,950

Price	Year	Aircraft
11,000	50	Navion A
11,500	58	Navion D
12,000	60	Mooney 20 A
13,000	60	Navion F
13,500	60	Bellanca 14-19-3
13,750	58	Lake C-2 IV
13,750	48	Beech Bonanza 35
14,500	68	Waco Sirius
14,500	60	Lake LA-4

$15,000–$19,950

Price	Year	Aircraft
15,500	62	Mooney 21
15,750	64	Lake LA-4
16,000	50	Beech Bonanza B 35
16,500	60	Cessna 210
16,750	64	Mooney 21
16,900	66	Lake LA-4
17,000	52	Beech Bonanza D 35
17,500	64	Bellanca 14-19-3A
18,000	66	Bellanca 14-19-3AB
18,000	66	Mooney Super 21
18,250	68	Lake LA-4
18,500	64	Mooney Super 21
18,750	68	Bellanca 260C
19,750	62	Cessna 210

$20,000–29,950

Price	Year	Aircraft
20,500	70	Piper Cherokee Arrow 180
22,500	68	Navion Rangemaster
22,500	62	Piper Comanche 250
22,500	56	Beech Bonanza G 35
23,000	70	Bellanca Viking
23,000	70	Piper Cherokee Arrow 200
23,500	72	Cessna Cardinal RG

The 1957 Bonanza sells for about $25,000 today—more than its original price! (Beech photo)

24,000	70	Bellanca Super Viking
24,750	60	Beech Debonair
25,000	70	Lake Buccaneer
25,000	64	Cessna 210
25,500	74	Beech Sierra
25,750	72	Bellanca Viking
26,500	72	Bellanca Super Viking
26,500	62	Beech Debonair
26,500	58	Beech Bonanza J 35
27,500	74	Cessna Cardinal RG
27,500	70	Bellanca Turbo Viking
27,500	70	Navion Rangemaster
27,500	62	Meyers 200 B
28,250	72	Lake Buccaneer
28,500	66	Cessna Centurion
28,500	64	Beech Debonair
28,500	64	Meyers 200 C
28,500	60	Beech Bonanza M 35
29,500	76	Beech Sierra

$30,000–$39,950

30,000	74	Piper Cherokee Arrow II
31,000	74	Lake Buccaneer
31,500	68	Cessna Centurion
32,000	76	Cessna Cardinal RG
32,000	72	Bellanca Turbo Viking
32,000	66	Piper Comanche 260
33,000	66	Meyers 200 D
33,000	62	Beech Bonanza P 35
32,500	76	Rockwell International 112 TC
33,500	66	Beech Debonair
33,750	74	Bellanca Viking
34,000	76	Lake Buccaneer
34,000	76	Piper Cherokee Arrow II
35,000	74	Bellanca Viking
35,000	68	Cessna Turbo Centurion
35,000	68	Piper Comanche 260
35,750	70	Cessna Centurion
36,000	68	Beech Bonanza E 33
39,000	70	Beech Bonanza F 33
39,000	64	Piper Comanche 400
39,500	64	Beech Bonanza S 35

A 1974 Piper Cherokee Arrow II, a popular used aircraft at about $30,000. (Piper photo)

$40,000–$49,950

40,000	78	Mooney Ranger
40,000	74	Bellanca Viking TC
40,500	70	Cessna Turbo Centurion
41,000	70	Piper Comanche 260
42,000	72	Beech Bonanza 33
42,000	72	Cessna Centurion
42,000	68	Mooney 22 Mustang
42,500	76	Rockwell International 112 TC
42,500	66	Beech Bonanza V 35
43,000	70	Piper Turbo Comanche
43,000	68	Beech Bonanza E 33C
43,250	78	Cessna 177 Cardinal RG
44,000	72	Piper Comanche 260
44,000	68	Beech Bonanza E 33A
44,500	72	Cessna Turbo Centurion
45,000	78	Lake Buccaneer
45,000	76	Rockwell International 114
45,000	76	Navion Rangemaster
46,000	72	Piper Turbo Comanche
47,500	78	Piper Arrow III

$50,000–$74,950

50,000	78	Piper Turbo Arrow III
52,000	78	Mooney 201
53,250	70	Beech Bonanza V 35 B TC
54,000	76	Piper Lance
54,000	70	Beech Bonanza A 36
54,750	76	Bellanca Turbo Viking

55,000	72	Beech Bonanza F 33A
57,500	72	Beech Bonanza V 35B
58,000	78	Bellanca Viking 300
58,000	76	Cessna Centurion
59,750	78	Bellanca Viking 31A-300
60,000	72	Beech Bonanza A 36
62,000	74	Beech Bonanza F 33A
64,000	74	Beech Bonanza V 35B
64,500	76	Cessna Turbo Centurion
65,000	78	Rockwell International 112 TC
67,500	78	Bellanca Turbo Viking
67,500	78	Rockwell International 114
68,750	78	Piper Lance II
69,000	74	Beech Bonanza A 36
69,500	78	Cessna Centurion
70,000	76	Beech Bonanza 33A

$75,000–$99,950

75,000	76	Beech Bonanza V 35B
76,000	78	Piper Turbo Lance II
80,000	78	Cessna Turbo Centurion
83,250	76	Beech Bonanza A 36
90,000	78	Beech Bonanza F 33A
91,250	78	Beech Bonanza V 35B
95,000	78	Beech Bonanza A 36

Over $100,000

105,000	78	Cessna Pressurized Centurion

■ TWIN ENGINE

Under $20,000

14,250	54	Piper Apache
15,250	56	Piper Apache
16,500	58	Piper Apache
17,000	64	Cessna 336 Skymaster
18,000	56	Cessna 310
19,000	60	Piper Apache

$20,000–29,950

22,500	52	Beech Twin Bonanza
23,000	66	Cessna Skymaster
24,900	60	Cessna 310
26,000	54	Beech Twin Bonanza
27,500	62	Piper Apache 235
28,500	68	Cessna Skymaster
28,500	62	Cessna 310
28,500	62	Piper Aztec B
28,500	58	Beech Travel Air
28,500	56	Beech Twin Bonanza
29,000	64	Piper Apache 235
29,500	64	Piper Twin Comanche

$30,000–39,950

30,000	64	Piper Aztec B
31,000	70	Cessna Skymaster
31,500	62	Cessna 320
32,500	68	Cessna Turbo Skymaster
33,250	60	Beech Travel Air
33,500	64	Cessna 310
34,500	72	Cessna 337 Skymaster
34,500	66	Piper Twin Comanche
35,000	58	Beech Twin Bonanza
36,500	70	Cessna 337 E Turbo Skymaster
37,250	62	Beech Baron A 55
37,500	72	Piper Seneca
37,500	68	Piper Twin Comanche
38,000	66	Piper Turbo Twin Comanche
38,500	66	Piper Aztec C
39,000	66	Cessna 310 K

$40,000–$49,950

40,000	64	Cessna 320B
40,000	60	Beech Twin Bonanza H 50
41,000	64	Beech Travel Air
41,500	66	Piper Turbo Aztec C
41,750	64	Beech Baron B 55
42,000	68	Piper Turbo Twin Comanche
42,500	74	Cessna Skymaster
42,500	60	Beech Twin Bonanza
43,000	66	Beech Travel Air D 95A
43,500	68	Piper Aztec C

A 1976 Cessna Turbo Centurion will set you back about $65,000. (Cessna photo)

44,500	68	Cessna 310N
45,000	74	Piper Seneca
47,500	70	Piper Twin Comanche
47,500	68	Piper Turbo Aztec C

$50,000–$74,950

51,000	70	Piper Turbo Twin Comanche
51,750	72	Piper Twin Comanche
52,500	70	Cessna 310Q

A late-model Piper Twin Comanche still outperforms the new light twins, at half the price.

52,500	70	Piper Aztec D
54,000	72	Piper Turbo Twin Comanche
54,000	68	Beech Baron B 55
56,500	68	Cessna 320F
57,500	70	Beech Baron B 55
58,000	70	Piper Turbo Aztec D
59,500	68	Beech Baron D 55
62,500	70	Cessna 310Q Turbo
63,000	72	Cessna 310Q
64,500	76	Cessna Skymaster
64,500	74	Cessna Pressurized Skymaster
65,000	72	Piper Aztec E
65,000	68	Beech Baron 56 TC
67,000	74	Cessna 310
69,500	72	Beech Baron B 55
69,500	70	Beech Baron 55
70,000	72	Piper Turbo Aztec E
72,500	70	Beech Baron 56TC

$75,000–$99,950

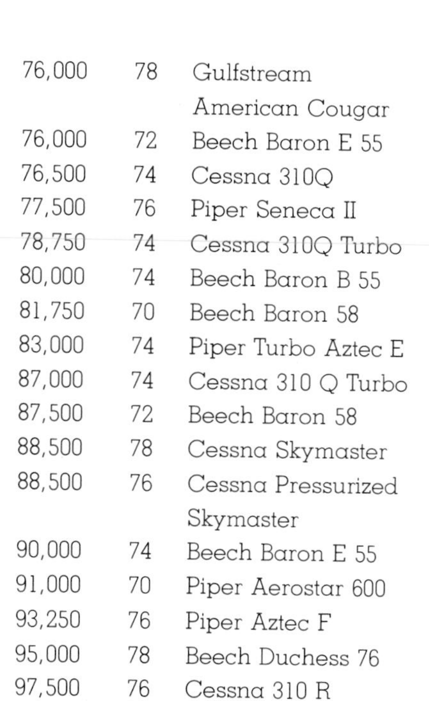

75,700	74	Piper Aztec E
76,000	78	Gulfstream American Cougar
76,000	72	Beech Baron E 55
76,500	74	Cessna 310Q
77,500	76	Piper Seneca II
78,750	74	Cessna 310Q Turbo
80,000	74	Beech Baron B 55
81,750	70	Beech Baron 58
83,000	74	Piper Turbo Aztec E
87,000	74	Cessna 310 Q Turbo
87,500	72	Beech Baron 58
88,500	78	Cessna Skymaster
88,500	76	Cessna Pressurized Skymaster
90,000	74	Beech Baron E 55
91,000	70	Piper Aerostar 600
93,250	76	Piper Aztec F
95,000	78	Beech Duchess 76
97,500	76	Cessna 310 R

The 1976 Cessna 310 goes for a little under $100,000. (Cessna photo)

$100,000–$199,950

100,000	76	Piper Turbo Aztec F
101,500	78	Piper Seneca II
102,500	78	Cessna Turbo Skymaster
103,000	70	Piper Aerostar 601 Turbo
105,000	76	Beech Baron B 55
107,500	74	Beech Baron 58
110,000	76	Cessna 310 R Turbo
110,000	68	Beech Duke 60
112,500	72	Cessna 340
115,000	76	Beech Baron E 55
117,500	74	Piper Aerostar 600 A

This Beech Duke will fetch a high price. (Beech photo)

122,500	74	Cessna 340
125,000	70	Beech Duke
132,500	78	Piper Aztec F
132,500	76	Piper Aerostar 600A
132,500	74	Piper Aerostar 601A Turbo
133,500	78	Cessna 310 R
135,000	78	Cessna Pressurized Skymaster
136,000	78	Beech Baron B 55
140,000	76	Beech Baron 58
140,000	72	Beech Duke A 60
142,500	78	Piper Turbo Aztec F
147,500	78	Cessna 310 R Turbo
150,000	76	Piper Aerostar 601A Turbo
157,500	78	Beech Baron E 55
160,000	78	Piper Aerostar 600A
162,500	76	Cessna 340 A
162,500	74	Piper Aerostar 601P
165,000	76	Beech Baron 58TC
165,000	74	Beech Duke B 60
177,500	76	Beech Baron 58P
180,000	78	Piper Aerostar 601B Turbo
182,500	76	Piper Aerostar 601P
185,000	78	Beech Baron 58

Over $200,000

200,000	78	Cessna 340 A
200,000	78	Beech Baron 58TC
210,000	76	Beech Duke B 60
215,000	78	Piper Aerostar 601P
250,000	78	Beech Baron 58P
300,000	78	Beech Duke B 60

P A R T T H R E E

Aircraft Operation

BUYING A PLANE
FINANCING AN AIRPLANE
INSURANCE
FIXED BASE OPERATORS
AVIONICS
ENGINES
FUEL
OIL
MODIFICATIONS
HANGARS
INTERIORS
PAINT
COMPLAINTS

The Rockwell Commander 112. (Rockwell photo)

Buying a Plane

I guess every pilot dreams of one day owning a plane. This is not as difficult as it may seem. All it takes is money, and lots of it! However, if you go out to your local small airport and look at the people who are untying their aircraft for some weekend flying, you won't find a lot of millionaires. You'll find plumbers and firemen and salespeople and factory workers, as well as the more obvious types like doctors and dentists. How do they do it?

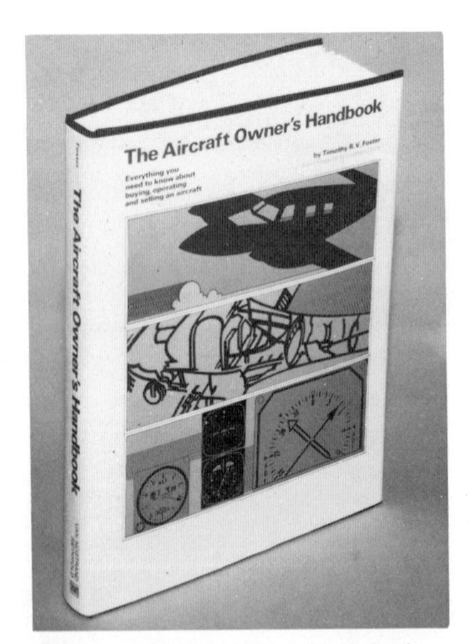

Strongly recommended by the author! (Tim Foster photo)

My book *The Aircraft Owner's Handbook—Everything You Need to Know About Buying, Operating and Selling an Aircraft,* published by Van Nostrand Reinhold Company, will prove very helpful in this area. It takes you through every aspect of aircraft ownership from considering it, funding it, and choosing it through owning it, operating it, fixing it, knowing it, and selling it.

Aircraft ownership may be divided into two broad categories—people who fly their aircraft for business reasons (and thus get a tax write-off) and people who fly for pleasure (and thus pay for it all themselves). Flying is so expensive that the first thing you should look for is some kind of business use for your plane. This can be difficult for a salaried employee, working in an office or a plant. But it's not impossible. You don't need a commercial pilot's license to fly an aircraft for business—a private certificate will do. You only need a commercial certificate if you are carrying passengers or freight for hire, or using the aircraft for aerial work, such as crop dusting or fire fighting.

Business use of aircraft

If you have a business use for an aircraft, all you have to do is determine which aircraft is best for the job and which is the best way to handle the financing, and get on with it. What operations constitute business use? Well, an aircraft is primarily a means of transportation, so if you have to travel on business, and you can be reimbursed or claim the travel expense, there's a business use for an aircraft. Bear in mind that you'll also need some form of ground transportation at the other end. This can consist of using rental cars, taxis, or public transportation, or getting the person you're going to see to come out to the airport to meet you. This latter is often quite easy to accomplish, and it impresses your clients no end!

The Bickerton folding bike—cheap ground transportation. (Aircraft Components photo)

If you're really adventurous, you can even get a folding bicycle and carry it on your airplane with you. One such model is the Bickerton Portable Bicycle, which weighs 22 lbs, folds up to a 30" × 20" × 9" package, and costs about $350. It has a 3-speed transmission, and may be ordered from **Aircraft Components, Inc., 700 North Shore Drive, Benton Harbor, MI 49022, telephone (800) 253-7261, (616) 925-8661 in MI.**

Another way to use an aircraft for business is to take aerial photographs and sell them. There's a book that tells you how—*Profits in Aerial Photography,* which costs $3.95 and may be ordered from **Roberts, 125 University Drive, East Hartford, CT 06108.**

The leaseback

One of the best business uses for an aircraft is the leaseback. With this method, you buy a suitable aircraft and lease it to someone who can put a fair amount of time on it. This puts you in the airplane-leasing business, and makes all the expenses associated with the operation of the aircraft while it is under lease tax deductible, with the leasing revenues being added to your other taxable income. This method is only attractive if you are in a high tax bracket—at least a 40 to 50 percent marginal tax rate. You can buy a trainer and lease it back to a flight school, or perhaps a light single or twin and lease it back to an air taxi/charter operator, or even buy an aircraft for your corporation and lease it back to them. See the box for a cash-flow analysis for just such a corporate leaseback, which assumes the purchase of a $100,000 airplane, a six-year lease, and a rental rate of $75 an hour, increasing by $5 per hour each year. We asssume 300 hours per year of utilization. If you're in a 50 percent tax bracket, you could end up owning the airplane at the end of six years for an actual net cash outlay of under $12,000.

The principal benefit available to a person buying an aircraft for business use is the *investment tax credit*. This credit can be 10 percent of the value of the aircraft (10 percent of the first $100,000 if it is a used plane), and it is a credit against the tax you pay, not a deduction from taxable income. Thus a $100,000 business aircraft purchase would reduce your tax bill by $10,000 in the first year. In addition to this, you have *depreciation*. This is a deductible expense designed to compensate you for the declining balance of the asset over its lifetime. You can use *accelerated depreciation* in some cases. See the box for an example. Talk to an accountant or tax attorney for the best advice on these approaches. Depreciation, while it is a business expense, is not a cash outlay. Thus, on a cashflow basis, you get a reduction in your tax payable due to a *bookkeeping entry* rather than a *cash expenditure*. This can result in a net increase in the cash you have available to you that year.

The Piper Archer is a popular leaseback vehicle. (Piper photo)

Using the manufacturer's resources

The manufacturers will use their own computer power to carry out a customized cash-flow analysis for you if you're thinking of buying a plane. Beech offers a Capital Recovery Guide and a business-flying kit **(Beech Aircraft Corporation, Wichita, KS 67201, telephone (316) 681-7111).** Cessna offers its Transportation Analysis Plan **(Cessna Aircraft Company, Wichita, KS 67201, telephone (316) 685-9111).** Piper offers a Cash Flow Analysis **(Piper Aircraft Corporation, Lock Haven, PA 17745, telephone (717) 748-6711).**

Buying an airplane for pleasure

I hope you're rich. Otherwise, consider a flying club (see page 21) or taking on a partner. Having a partner can be very helpful, because the most expensive part of owning a plane is the fixed costs—hangarage or tiedown, insurance, annual inspection, interest expense, and so on. The direct operating costs—fuel and oil, engine overhaul reserve, and such represent a smaller portion of the costs. You can obtain a partner by advertising in your local paper, putting a sign up at your airport bulletin board, or asking your aircraft dealer to make a connection (it's in his interest to do this, since he might sell an airplane to you). Look at the tables of used airplane prices on pages 105–111 to get an idea of what one may cost. Yes, there are airplanes available for under $5,000. Don't forget my book *The Aircraft Owner's Handbook*. It's full of information on buying an airplane.

Annual Expense and Income Table for Aircraft on Leaseback

	YEAR					
	1	2	3	4	5	6
Estimated cash outflow before tax						
Down payment	20,000					
Principal	13,333	13,333	13,333	13,333	13,333	13,333
Interest*†	5,694	5,694	5,694	5,694	5,694	5,694
Fuel and oil*	4,000	4,400	4,840	5,324	5,856	6,442
Insurance and hangar*	3,000	3,000	3,000	3,000	3,000	3,000
Maintenance*	1,500	1,500	1,500	1,500	1,500	1,500
Cash out before income	47,527	27,927	28,367	28,851	29,383	29,969
Lease income	22,500	24,000	25,500	27,000	28,500	30,000
Deductible expenses*	−14,194	−14,594	−15,034	−15,518	−16,050	−16,636
Gross business profit	8,306	9,406	10,466	11,482	12,450	13,364
Less Depreciation*	36,000	21,333	14,222	9,842	6,321	4,214
Taxable profit (loss)	(27,694)	(11,927)	(3,756)	(1,640)	(6,129)	(9,150)
[*increase (reduce) your personal taxable income by this amount*]						
Tax result (50% rate)	−13,847	−5,964	−1,878	+820	+3,065	+4,575
Investment tax credit	−10,000					
Total tax result	−23,847	−5,964	−1,878	+820	+3,065	+4,575
[− = *tax saving to you;* + = *tax due from you*]						
Cash out after income	25,027	3,927	2,867	1,851	883	−31
Cash out (in) after tax	1,180	(2,037)	989	2,671	3,948	4,544

Total cash outflow after taxes, over six years: $11,295

*Deductible expense
†Note: if interest tax deduction is amortized, results will vary slightly

Assumptions:
Initial cost $100,000
Aircraft financed for 6 years
Finance rate 12.5% APR, simple interest
300 hours flown per year (about 45,000 miles)
Fuel and oil cost increased 10% per year
Income tax rate 50%, joint return filed
Double-declining balance depreciation used, over 6 years
Aircraft leased at $75 per flight hour, rate increased
$5 per hour per year to compensate for rising costs
No reserve for engine overhaul, due after year 6

Beech Baron 58. (Beech photo)

Mooney 201. (Mooney photo)

Piper Turbo Aztec F. (Piper photo)

Cessna 310. (Cessna photo)

Piper Seneca II. (Piper photo)

Depreciation of $100,000 Airplane Over Six Years—Double-Declining Balance

With a useful life of six years or more, you can take a bonus depreciation of 20% of the first $10,000 of its cost—$2,000. If you file a joint return, you can take $4,000. This leaves a cost basis of $96,000. Let's assume you depreciate the aircraft over six years, using the double-declining balance method, and that you file a joint return. Here's how the annual depreciation looks:

Year	Starting Value	Depreciation	Ending Value
1	$96,000	$32,000	$64,000
2	$64,000	$21,333	$42,667
3	$42,667	$14,222	$28,445
4	$28,445	$9,482	$18,963
5	$18,963	$6,321	$12,642
6	$12,642	$4,214	$8,428

Beech Bonanza A36. (Beech photo)

Cessna Skyhawk. (Cessna photo)

Why the A36 is popular. (Beech photo)

Financing an Airplane

Airplanes cost a lot of money, so they often take specialized financing. Your friendly local bank may not be the best place to do this for you—they probably think airplanes are dangerous, and they don't understand the relative values of such things as avionics, high- vs low-time engines, and so on. Fortunately there are several banks and institutions that specialize in aircraft. The Aircraft Finance Association (Box 595, Wichita, KS 67201) lists these among their membership:

California

Crocker National Bank
74 New Montgomery Street
San Francisco, CA 94105
(415) 983-6613

Golden State Sanwa Bank
220 Almaden Boulevard
San Jose, CA 95113
(408) 998-0800

The Bank of Montecito
Box 5010
Santa Barbara, CA 93108
(805) 969-5881

Security Pacific National Bank
Aircraft Division
Box 2638
Van Nuys, CA 91404
(213) 768-4222

Colorado

AeroBankers, Inc.
3333 Quebec Street, Suite 9125
Denver, CO 80207
(303) 320-1977

First National Bank of Denver
Box 5808
Denver, CO 80217
(303) 893-2211

Connecticut

Citytrust
961 Main
Bridgeport, CT 06602
(203) 384-5552

First New England Financial Corporation
411 Pequot Avenue
Southport, CT 06490
(203) 255-5713

General Electric Credit Corporation
Box 8300
Stamford, CT 06904
(203) 357-4745

Delaware

National Financing, Inc.
Box 1922
Wilmington, DE 19899
(302) 322-7385

Florida

First State Bank of Miami
Box 458
Hialeah, FL 33011
(305) 823-1270

Piper Acceptance Corporation
Box 970
Lakeland, FL 33802
(813) 644-3558

Citizens Bank of Palm Beach County
4395 Southern Boulevard
West Palm Beach, FL 33406
(305) 683-6800

Georgia

Fulton National Bank
Box 4387
Atlanta, GA 30302
(404) 529-4540

Grumman Credit Corporation
Box 2206
Savannah, GA 31402
(912) 964-8506

Illinois

Beverly Bank
1357 W 103 Street
Chicago, IL 60643
(312) 881-2369

Harris Trust and Savings Bank
111 West Monroe Street
Chicago, IL 60690
(312) 461-5115

Kansas

Johnson County National Bank
6940 Mission Road
Prairie Village, KS 66208
(913) 362-7000

Beech Acceptance Corporation, Inc.
Box 85
Wichita, KS 67201
(316) 681-7164

Cessna Finance Corporation
Box 308
Wichita, KS 67201
(316) 685-5456

Fourth National Bank and Trust Company
100 North Broadway
Box 1318
Wichita, KS 67201
(316) 261-4468

Maryland

Commercial Credit Equipment Corporation
Aircraft Division
300 St Paul Place
Baltimore, MD 21202
(301) 332-3626

Maryland National Bank
Box 535
Baltimore, MD 21203
(301) 244-5544

Massachusetts

Attleboro Trust Company
Box 330
Attleboro, MA 02703
(617) 222-6100

General Discount Corporation
100 State Street
Boston, MA 02109
(617) 277-0900

South Shore Bank
MultiBank Aviation
Norwood, MA 02062
(617) 769-5238

Michigan

Michigan National Bank of Detroit
1250 West 14 Mile Road
Clawson, MI 48017
(800) 521-7188

Minnesota

American National Bank and Trust Company
370 Minnesota Street
St. Paul, MN 55101
(612) 298-6356

New Jersey

United Jersey Bank
210 Main Street
Hackensack, NJ 07602
(201) 646-5175

Yegen Air Acceptance Corporation
365 West Passaic Street
Rochelle Park, NJ 07662
(201) 368-9600

National Community Bank
24 Park Avenue
Rutherford, NJ 07070
(201) 939-1027

New York

Manufacturers and Traders Trust Company
Box 767
Buffalo, NY 14240
(716) 626-3308

Bankers Trust Company
280 Park Avenue
New York, NY 10017
(212) 692-2040

National Bank of North America
44 Wall Street
New York, NY 10005
(212) 623-4076

Ohio

Central Trust Company of NE Ohio
Box 9280
Canton, OH 44711
(216) 455-6711

Central National Bank of Cleveland
Consumer Credit Dept., 7th floor
800 Superior Avenue
Cleveland, OH 44114
(216) 861-7800

Park Leasing Corporation
50 North Third Street
Newark, OH 43055
(614) 349-8451

Oklahoma

Rockwell International Credit Corporation
5001 North Rockwell Avenue
Bethany, OK 73008
(405) 787-6000

Pennsylvania

First Pennsylvania Bank N.A.
1500 Chestnut Street
Philadelphia, PA 19101

National Leasing Corporation
3420 Mellon Bank Building
Pittsburgh, PA 15219
(412) 281-7300

American Bank and Trust Company of Pennsylvania
Box 189
Reading, PA 19603
(215) 375-5338

Tennessee

Bank of Hendersonville
Box 8
Hendersonville, TN 37075
(615) 824-6542

Union Planters National Bank of Memphis
Aircraft Finance Department
Box 387
Memphis, TN 38147
(901) 523-6446

Texas

National Bank of Commerce
Box 3190
Brownsville, TX 78520
(512) 546-2292

Avemco Aircraft Investment Corporation
Box 4216
Fort Worth, TX 76106
(817) 626-5444

Virginia

United Virginia Leasing Corporation
Box 189
Richmond, VA 23202
(804) 782-5458

Insurance

Aviation insurance is offered through *direct underwriters, agents,* and *brokers.* The underwriter is the firm that actually carries the insurance and that pays you in the event of a claim. Some underwriters deal directly with the public, others deal through brokers or agents. A broker is an agent who shops around for you and finds the best deal, according to your own special needs. An agent represents one or more specific insurance companies.

One of the most widely known direct underwriters is **Avemco Insurance Company, Box 30007, Bethesda, MD 20014.** They have a variety of telephone numbers, depending on where you are calling from. If you live in Maryland, call collect to **(301) 986-4700.** If you live otherwise east of the Mississippi, call **(800) 638-8440.** If you live in Texas, call **(800) 792-1261.** If you live otherwise west of the Mississippi, call **(800) 433-1750.** Avemco claims to write more insurance policies on pilots, aircraft, and flying clubs than any other insurance company in the world. Part of this is attributable to their long-standing endorsement by the Aircraft Owners & Pilots Association (AOPA). If you belong to AOPA, you automatically get an Avemco group life policy, and since AOPA has over 230,000 members, it's easy for Avemco to make that claim.

Another direct underwriter is **National Aviation Underwriters, Box 10155, St Louis, MO 63172, telephone (800) 325-8988.** They operate through regional agents, so if you call the 800 number, or **(314) 426-4000** if you live in Missouri, they'll put you in touch with your local agent.

Among brokers specializing in aviation insurance are:

Alexander & Alexander, Inc.
1185 Avenue of the Americas
New York, NY 10036
(212) 575-8000

Bayly, Martin & Fay Aviation Insurance Services, Inc.
3200 Wilshire Boulevard
Los Angeles, CA 90010
(213) 381-5371

Bayly, Martin & Fay Aviation Insurance Services, Inc.
1700 Market Street
Philadelphia, PA 19103
(215) 561-5700

Richard J. Berlow & Company, Inc.
Box 1968
Teterboro, NJ 07608
(800) 631-1952; (201) 288-1091 in NJ

Corroon & Black Corporation
1700 Market Street
Philadelphia, PA 19103
(215) 568-5505

Don Flower Associates Inc.
Box 1482
Wichita, KS 67201
(316) 943-9333

Frank B. Hall & Company
261 Madison Avenue
New York, NY 10016
(212) 682-7200

Marsh & McLennan, Inc.
1221 Avenue of the Americas
New York, NY 10020
(212) 997-2000

Among underwriters your broker might cover you through are:

Associated Aviation Underwriters
90 John Street
New York, NY 10038
(212) 766-1600

Aviation Office of America
Love Field Terminal Building
Dallas, TX 75235
(214) 350-8911

Insurance Company of North America
1600 Arch Street
Philadelphia, PA 19103
(215) 241- 3875

Southern Marine and Aviation Underwriters
601 Poydras Street
New Orleans, LA 70130
(504) 524-4131

United States Aviation Insurance Group
110 William Street
New York, NY 10038
(212) 349-2100

Why you need insurance. (Chris Foster photo)

Types of coverage available

You can insure almost anything these days. If you own a plane, you can insure the bird itself (hull coverage) to protect you against loss caused by accident or damage, usually with a $250 to $1,000 deductible; if you own or rent a plane, you can insure yourself against a liability claim arising out of your use of the aircraft; if you own a hangar, you can insure that, and so on. See the section on Learning to Fly, page 113 for more information on the insurance aspects of taking lessons.

Fixed Base Operators

The FBO provides all or part of these functions:

- New and used aircraft sales
- Avionics sales
- Parts sales
- Maintenance and service
- Storage
- Fuel
- Flight training
- Aircraft rentals
- Air-taxi and charter services
- Aircraft management services

Some FBOs are a delight to visit, and some are not. Every year *Professional Pilot* magazine conducts a survey among its readers (who spend a lot of time flying all over the country from one FBO to another) to see which is the best. Here, reprinted with permission, is the *Professional Pilot* list of the top thirty FBOs in the United States in 1979.

"Our eager staff is ready to serve you!" (Beech photo)

The Top 30 F.B.O.s (1979)

Rank	Operator Name	Location	State,	Airport
1	Combs Gates	Denver	CO	DEN
2	Flower Aviation	Pueblo	CO	PUB
3	Memphis Aero	Memphis	TN	MEM
4	Combs Gates	Indianapolis	IN	IND
5	Midcoast Aviation	St. Louis	MO	STL
6	Nashville Jet Center	Nashville	TN	BNA
7	Flower Aviation	Salina	KS	SLN
8	Cooper Airmotive	Dallas	TX	DAL
9	Aero Facilities	Miami	FL	MIA
10	Beckett Tilford	W. Palm Beach	FL	PBI
11	Walkers Cay	Ft. Lauderdale	FL	FLL
12	Combs Gates	Palm Springs	CA	PSP
13	Executive Beech	Kansas City	MO	MKC
14	Hangar One	Atlanta	GA	ATL
15	Duncan Aviation	Lincoln	NE	LNK
16	Exec Air	Grand Island	NE	GRI
17	General Aviation Corp	New Orleans	LA	MSY
18	Denver Beechcraft	Denver	CO	DEN
19	Stevens Beechcraft	Greer	SC	GSP
20	Page Airways	Washington	DC	DCA
21	Aero Services	Teterboro	NJ	TEB
22	Atlantic Aero	Greensboro	SC	GSO
23	AiResearch	Los Angeles	CA	LAX
24	Showalter Flying Service	Orlando	FL	ORL
25	Interair	St. Petersburg	FL	PIE
26	Casper Air Service	Corpus Christi	TX	CRP
27	Atlantic Aviation	Hobby	TX	HUB
28	Mitchell Aero	Milwaukee	WI	MKE
29	Hangar One	Birmingham	AL	BHM
30	Hughes Aviation Service	Las Vegas	NV	LAS

Avionics

Aviation electronics—avionics—are almost as important to an aircraft as its engine, and often more expensive. Some people regard an airplane as nothing more than a device for transporting a sophisticated array of avionics around the country. Today's avionics appeal to even the most ardent gadget freak, what with their digital readouts, pushbuttons, automatic features, memory devices, and logic patterns. And with the increasing restrictions and controls placed upon flight operations within different bits of airspace throughout the world, more and more avionics are needed in aircraft. Avionics requirements in general aviation aircraft can be divided into four broad categories: communication, navigation, identification, and control. Most avionics manufacturers produce complete lines of components to cover all the basic needs, although some do not fill one particular need or another. Let's look at the manufacturers first, then at the specific devices.

Avionics manufacturers

Avionics manufacturers tend to *specialize* in avionics. You won't see the familiar home-electronics manufacturers represented, such as Zenith, Sony, General Electric, Pioneer, Panasonic, and so on. One exception is RCA, which makes some components, but not a full line. Avionics producers tend to address specific market segments, although a few attempt to serve the whole industry. The market segments may be categorized as military, airline, corporate, medium, light, and light-retrofit. Military, airline, and corporate avionics are beyond the scope of *The Aviator's Catalog*. The other three categories are served by most avionics manufacturers, as you will see.

The "big four" avionics manufacturers are Cessna, Collins, King, and Narco. These are the brands most often factory installed in new aircraft. Another strong contender is Bendix, which offers a total line of radios priced to appeal more to the medium-level and corporate users. Two other manufacturers, Edo-Aire and Genave, seem to serve mostly the retrofit market, since you don't see too many of their products being offered by the aircraft manufacturers.

Cessna is the only high-volume aircraft manufacturer that has its own line of avionics, and these are standard or optional equipment in various Cessna models. Cessna bought out the old Aircraft Radio Corporation some years ago, changed the "ARC" to Aircraft Radio and Controls division of Cessna Aircraft Company, and developed a complete line of avionics. Cessna does not aggressively market its radios outside of their aircraft, so you seldom see them except in Cessnas, and you don't usually buy them except already installed in an airplane. Cessna avionics come in three basic series, 300, 400, and 800/1000. The 300 series is the lowest price line, while the 800/1000 series is the top of the line. **Cessna Aircraft Company, Box 1521, Wichita, KS 67201, telephone (316) 685-9111.**

The Cadillac of the avionics industry has always been Collins, which is now a subsidiary of Rockwell International. Rockwell also owns Aero Commander and the old North American Aircraft, so they, too, are in the airplane-building business, but they don't seem to be so insistent about putting Collins radios into Rockwell airplanes (it is very hard to get Cessna to put a non-Cessna radio in one of its planes). Rockwell International aircraft are available with radios by a variety of manufacturers. Collins makes radios that are the top of the line, and these will be found in most airlines and business jets. In recent years, however, they have recognized the existence of the smaller general-aviation airplanes and introduced the Microline, which is a complete pack-

Would you believe this is a single-engine airplane? (Mooney photo)

age of lightweight, low-power-drain radios that can now be found even in trainers, such as the Piper Tomahawk. A few years ago it was unheard of to see Collins Radios in such aircraft! Collins also offers the Pro Line, designed to appeal to the heavier twins and corporate-aircraft users. Collins radios are sold through factory-appointed dealers and as optional original equipment in most new aircraft being offered today. **Collins General Aviation Division, Rockwell International, Cedar Rapids, IA 52406, telephone (319) 395-1000.**

Ed King used to work for Collins, and started King Radio Corporation in 1959. He aimed initially at the general aviation market—the first radio I remember was the KY 90, which was a 90-channel communications radio (COM) available when most other manufacturers were offering tunable receivers or 8-channel sets. King has always been an innovator, which is particularly noticeable today with many of the radios in their lines. They offer two basic lines—the Gold Crown and Silver Crown, although now there is a dichotomy in Silver Crown, since some of the older, more mechanical sets are being offered side-by-side with new digital-readout, microprocessor-controlled units with all kinds of extra features. This has lead some wags to refer to the cheaper Silver Crown models as the "Brown Crown" line! From their start, King has made dramatic inroads at the top end of the line and now sells much equipment to airlines and top corporate aircraft. King sells through aircraft manufacturers in factory installations on new aircraft and through their own network of dealers. **King Radio Corporation, 400 North Rogers Road, Olathe, KS 66061, telephone (913) 782-0400.**

Narco Avionics has for a long time been the "standard brand" light-aircraft radio. Narco has never addressed the top end of the line, although it has tried. In 1979, it introduced a series aimed at the medium level of aircraft users, the "E Line," but this was withdrawn soon after. Its main product offering is the "Centerline." This is a complete array of radios aimed at the new and retrofit aircraft market. Narco is also working on its first weather radar. The Narco Mark 12 NAV/COM was for ages the most popular general-aviation radio around, even though it has been out of production for years. However, this distinction has been taken over by the King KX 170/175 series NAV/COM. Narco sells through its own dealers and their products are offered by most general aviation manufacturers as optional equipment installed at the factory. **Narco Avionics, Fort Washington, PA 19034, telephone (215) 643-2900.**

The Bendix Avionics Division of the Bendix Corporation offers the BX 2000 line, an innovative package of radios unlike any others on the market. The radios are integrated with each other in unusual combinations and offer many features unavailable elsewhere. This line is substantially more expensive than its competition, but it is finding acceptance with many users who want the unique advantages Bendix offers. Some aircraft manufacturers offer the Bendix line as a factory-installed option, and it is available through Bendix dealers. **The Bendix Corporation, Avionics Division, Box 9414, Fort Lauderdale, FL 33310, telephone (305) 776-4100.**

Edo-Aire makes radios, autopilots, and flight-director systems. The two latter categories are available through some aircraft manufacturers, but their radios seem to sell only through dealers, so they're seldom seen in anything but retrofits. Edo-Aire also makes airplane floats and VOR ground stations. The autopilots are sold under the Mitchell brand name. **Edo-Aire Group Marketing Division, 216 Passaic Avenue, Fairfield, NJ 07006, telephone (201) 228-1800. Edo-Aire-Mitchell, Box 610, Mineral Wells, TX 76067, telephone (817) 325-2517.**

Genave offers a line of attractive, low-cost models for the retrofit market. One of the most interesting units is their Alpha Six hand-held 6-channel COM unit, suitable for use by people on the ground at airports, but also as an emergency unit for people who fly on instruments (IFR) and who want a backup radio that doesn't depend on their electrical system. They are represented by **Sales, Inc., telephone (800) 321-1188. General Aviation Electronics, Inc., 4141 Kingman Drive, Indianapolis, IN 46226, telephone (317) 546-1111.**

Sunair specializes in HF radios, used for long-distance flights over water and in the remote bush areas of the world. **Sunair Electronics, Inc., 3101 SW 3rd Avenue, Fort Lauderdale, FL 33315, telephone (305) 525-1505.**

Terra Corporation specializes in low-cost COM and NAV radios for the retrofit and homebuilt market. They service their products by mail. **Terra Corporation, 3520 Pan American Freeway NE, Albuquerque, NM 87107, telephone (505) 345-5621.**

The technical standard order (TSO)

The FAA has issued a set of specifications about environmental and performance capabilities for each type of radio equipment. The areas considered are:

- Temperature
- Humidity
- Vibration
- Audio-frequency susceptibility
- Radio-frequency susceptibility
- Spurious energy
- Explosion
- Electrical performance

For a piece of equipment to comply with the TSO, a sample must undergo certain tests, which set limits within which the equipment must operate. If it passes these tests, it may be allocated a TSO-compliance nameplate, which is to be found on the equipment.

A TSO'd component meets the limits set by the FAA for compliance. TSO compliance doesn't necessarily mean that non-TSO'd equipment is worse than TSO'd. It just means that if a product is TSO'd, it meets certain environmental and performance limits. Certain components, notably transponders, *must* be TSO'd.

Some manufacturers offer similar equipment in TSO'd and non-TSO'd versions, with a slight price saving in the non version. You pays your money, etc.

Communication radios (COM)

All civil aviation two-way communication in the populated parts of the world takes place in the VHF band between 118.000 and 135.975 mHz. In some countries, the 117.0 to 117.9 band is also used. Military air communication is in the UHF band. Long-range communication (such as is needed on transatlantic flights, or in the Arctic bush) is in the HF band, from 2 to 18 mHz.

The state of the art in VHF communications (COM) radios today is the 720-channel transceiver. In the past 20 years, general-aviation radios have gone from a tunable analog receiver, with maybe four or five transmitting crystals, through 90-channel simplex COMs (simplex means that the transmit and receive frequencies are the same), which operated between 118.0 and 126.9 mHz in 100-kHz steps, then to 360-channel sets (118.00 to 135.95 in 50-kHz steps), to the present 720-channel units (118.000 to 135.975 in 25-kHz steps).

What with microprocessors, solid-state design, and frequency synthesizers operating from a single crystal, today's COMs are a miracle of compactness, low weight, and low power drain.

Let's take a look at some of the COM units on the market. My favorite is the King KY 196/197 model, part of the Silver Crown line. I have one in my Comanche. This is a 720-channel unit with dual-frequency readout. The frequency on the left is the frequency in use, while the one on the right is the standby frequency. When you change frequencies, the change takes place in the standby position. When you want to use the newly selected frequency, you simply push the <—> button, and the two frequencies displayed change position, so that the one that was in the *use* position moves to the *standby* location, and vice versa. This is often referred to as a "flip-flop" feature. It is very useful, especially when you fly IFR, since you're constantly changing frequencies. Sometimes the new frequency doesn't answer. With the flip-flop button, you simply recall the old freq from the standby position and make a new deal with the controller. The KY 196 model is for 28-volt systems, and the 197 is for 14-volt systems. The unit is TSO'd, fully self-contained, weighs 3.2 pounds, and occupies a 6.25-inch-wide by 1.3-inch-high by 10.5-inch-deep space on your panel. It sells for about $1,800.

Narco makes the basic Centerline COM 120, a 720-channel

King KY 196 COM. (King photo)

Narco COM 120-20. (Narco photo)

TSO'd radio. The COM 120-20 is similar, and offers 20 watts of transmitting power. The COM 120 sells for about $1,300. The COM 120-20 costs just under $1,500. They are the same size: 6.25 inches wide (the standard avionics width for most radios, except Cessnas) by 1.75 inches high by 11 inches deep, and they weigh 3.5 pounds.

Collins makes two Microline COM units, the VHF-250 and VHF-251, both TSO'd. The 250 has mechanical frequency display, while the 251 has digital electronic display and offers storage of one extra frequency. However, the stored frequency is not displayed as it is on the King KY 196 and Narco COM 200. The Collins VHF-250 sells for a little over $1,000, while the 251 costs about $1,400. The Collins units are only half the width of standard avionics. This means that they can be stacked side-by-side with another COM or one of the Microline NAV units. The units are 3.12 inches wide by 2.61 inches high by 12.45 inches deep, and weigh 3.4 pounds.

Collins VHF-251 COM with digital tuning. (Collins photo)

Neither Cessna nor Bendix produce stand-alone COM units in the basic general-aviation range. Edo-Aire has several models and is one of the few manufacturers still to offer a 360-channel COM, the RT-551. The similar 720-channel version is the RT-551A. These radios are not-TSO'd. TSO'd versions are the RT-661 (360 channel) and RT-661A (720 channel). They are all 3.25 inches wide by 2.62 inches high by 11.44 inches deep, and weigh 3.3 pounds. They are all priced between $900 and $1,200. Stacking two side-by-side will put you 1/4 inch over the standard width (as found on King, Collins, and Narco), which could be a problem—the "wide" Edo-Aire radios are also 6.5 inches rather than the standard 6.25 inches in width. By the way, Cessna standard width is 6.625, giving a further incompatibility among them all. In other words you can mix and match Bendix, King, Collins, and Narco radios on the same stack, but you can't do this neatly with Edo-Aire and Cessna radios.

Collins VHF-250 COM with standard tuning. (Collins photo)

High frequency communications (HF COM)

You don't see too many HF sets in aircraft in the continental United States. HF is primarily a long-distance communications aid. It's used on transoceanic flights and in the bush. **Brelonix, Inc., 106 North 36th Street, Seattle, WA 98103, telephone (206) 633-1964,** offers several models, starting with the SAM-70 five-channel unit at about $1,600. Collins has the HF 200, which has 20 channels and costs about $8,000. **PanTronics Corporation (Box 22430, Fort Lauderdale, FL 33335, telephone (305) 525-4366)** offers several models, starting with the PT-10A, a ten-channel unit at about $1,800. The specialist in HF communications is **Sunair Electronics (3101 SW Third Avenue, Fort Lauderdale, FL 33315, telephone (305) 525-1515).** They offer seven different units, ranging from the six channel ASB-60, at $2,645, up to the 280,000 channel ASB-850, at almost $10,000.

NAV/COMS

Many COM units are offered in combination with a navigation radio, and thus are called NAV/COMs. Most manufacturers offer these. Let's take a look at some.

The most popular NAV/COM is the King KX 170B/175B series. Almost 100,000 of these have been sold. It consists of a 720-channel COM and a 200-channel NAV. The KX 170B is not TSO'd, while the 175B is. They will provide channeling for a remote DME, a

The best-selling King KX 175B NAV/COM. (King photo)

remote glideslope, and will drive an RMI (radio magnetic indicator) or HSI (horizontal situation indicator). They will also provide outputs for RNAV (area navigation). The KX 170B sells for about $1,900 and the 175B for about $2,000. Both measure 6.25 inches wide, 2.5 inches high, and 13 inches deep, and weigh in at 7 pounds. You need a course deviation indicator (CDI), HSI, or RMI to read the VOR signal. The King KI 203 is a straight VOR/ILS converter indicator, while the KI 204 presents a glideslope needle as well. The needle moves in a rectilinear presentation. If you prefer a lower cost model, you can select the KI 208 or KI 209, with needles that move as would a windshield wiper, pivoting about their fulcrums. These units sell for between $500 and $1000, depending on the model, and weigh between 1 and 1.5 pounds.

Cessna makes several NAV/COM models. The RT 385A is the basic 720-channel COM/200-channel NAV, using electronic digital readout of the frequencies. This sells for about $1,875. It measures 6.625 inches wide by 2.5 inches high by 11 inches deep, and weighs 5.2 pounds. To this must be added a CDI, such as the IN-386A (about $700), or the IN-386AC ($900), which includes a glideslope needle and automatic radial centering. The RT-485A is a more exotic NAV/COM, offering a 3-frequency memory on both the COM and NAV sides. A compatible CDI would be the IN-486AC model. TheNAV/COM costs about $3,000, while the indicator runs another $1,360 or so. It has the same dimensions as its cheaper sister above (the RT 385A), and weighs 5.5 pounds.

Narco doesn't make a NAV/COM in the Centerline series.

Cessna RT 385A NAV/COM. (Cessna photo)

Cessna 400 series NAV indicator. (Cessna photo)

Edo-Aire offers radios with full COM and simultaneous NAV capabilities. One version is the RT-553 (360 channels) or the RT-553A (720 channels). This throws in an extra 200 COM channels on the NAV tuner, so when you're not navigating, you can have the equivalent of a dual COM unit. However, whenever you are communicating, you have no NAV signal, making the radio strictly a VFR-type unit or possi-

Cessna RT 485A NAV/COM. (Cessna photo)

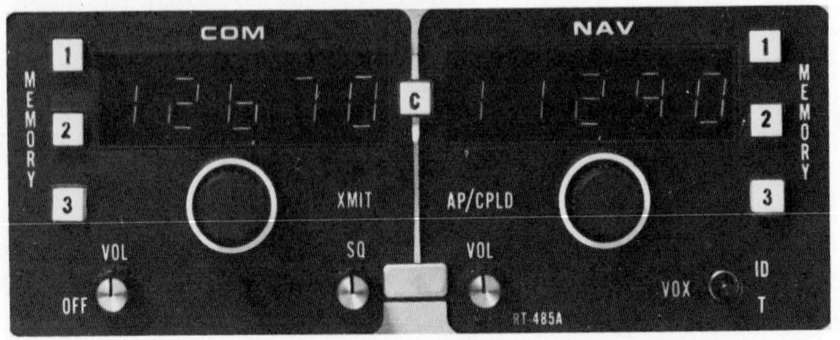

bly an IFR-backup system. The better version is the RT-563, which gives you a completely separate COM and NAV, with 360 channels (or 720 channels in the RT-563A), with a self-contained CDI. An additional feature on all of these models is automatic omni, which causes the VOR needle to center with a "TO" reading when a knob is pushed. This gives a fast track to the station, if you need it. All the units are the non-standard 6.5 inches wide by 3.25 inches high by 11.6 inches deep, and weigh between 5.5 and 7.8 pounds, depending on the model.

Genave has the GA 1000, a self-contained 720-channel COM and 200-channel NAV with built-in CDI. This unit is priced at less than $1,400 and weighs 5 pounds. It features the fashionable electronic digital readouts of frequencies.

Bendix has three TSO'd NAV/COMs, collectively called the CNA-2010 series, in its BX 2000 line. The CNA 2011A is a *dual* NAV/COM, containing two separate 720-channel COMs, two 200-channel NAVs, optional glideslope and marker receivers, and full audio switching circuitry. It features frequency preselect for the COM, gas discharge frequency displays, optional keyboard frequency selection, and built-in marker-beacon lights, among other things. It measures 6.25 inches wide by 4.5 inches high by 13 inches deep, weighs 13.5 pounds, and sells for about $5,300. The CNA 2012A is a single NAV/COM with frequency preselect on the COM and optional glideslope and marker receivers. It measures 6.25 inches

Bendix CN 2012A NAV/COM. (Bendix photo)

wide by 2.75 inches high by 13.5 inches deep, weighs 7.9 pounds, and sells for about $2,900. The CNA 2013A is similar to the CN 2012A, but does not include the frequency preselect feature. It has the same measurements, weighs 7.5 pounds, and costs about $2,000.

Bendix NAV/COMs can drive a Horizontal Situation Indicator (HSI) or Radio Magnetic Indicator (RMI), or their own unique electronic course deviation indicator (ECDI), which has no moving parts—instead of needles, it has light bars that move to indicate a displacement. The VOR radial, or "TO" bearing, is displayed digitally on the ECDI, and you can also have DME distance or

The Bendix ECDI shows ILS presentation (fly up and to the right). 090° is ILS course. 328° is missed approach radial from NAV 2. (Bendix photo)

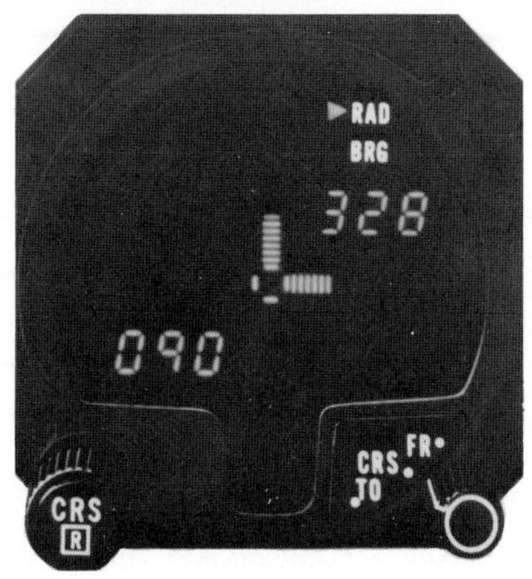

groundspeed displayed if you want. The Bendix ECDIs cost about $1,000 and fit a standard three-inch instrument hole.

VHF NAV receivers

If you don't get your navigation information from a NAV/COM, you need a straight NAV receiver and an instrument to read the steering information, such as a horizontal situation indicator (HSI), course deviation indicator (CDI) or radio magnetic indicator (RMI).

The most innovative NAV package on the market, in my opinion, is the Narco NAV 122. I have one in my Comanche. This provides, in one three-inch instrument hole, a CDI, a VHF 200-channel tuner, a VOR converter, an ILS glideslope receiver and needle, and a three-light marker beacon. The whole package weighs 3.3 pounds, is 10.75 inches deep, and costs just under $2,000. The same unit is available without glideslope and marker receivers, but with the

Narco's NAV 122 (left) and NAV 121 (right). (Narco photo)

glideslope needle, for about $1,700, as the NAV 122A. If you don't want glideslope presentation at all, get the NAV 121, which sells for about $1,250.

Narco NAV 124 tuner. (Narco photo)

Narco makes a separate TSO'd NAV tuner, the NAV 124, to stack with your other avionics; you would use this if you were driving an HSI or a RMI, or an existing CDI. It includes a glideslope receiver (NAV 124A without it). The unit measures 6.25 inches wide by 1.75 inches high by 11 inches deep, weighs 4.5 pounds, and sells for about $2,600 ($2,400 without glideslope).

King KN 53 NAV receiver. (King photo)

King offers a very nice TSO'd Silver Crown NAV receiver, the KN 53. This is similar in size and appearance to their KY 196 COM unit, and has the same frequency preselect feature. It includes a glideslope receiver and measures 6.25 inches wide by 1.3 inches high by 10.5 inches deep, weighs 4.7 pounds, and costs about $1,900. A KI 204 CDI costs about another $1,000.

Collins has the VIR 350 and 351, which are the same size and shape as the companion VHF COM units, the VHF 250 and 251. The Collins NAV 351 features

The Collins VIR-351 NAV receiver can display the VOR radial, as shown here. (Collins photo)

electronic frequency display, frequency storage (not displayed all the time), and digital VOR readout. The NAV 350 offers mechanical frequency display, with no storage or digital VOR. Both units measure 3.12 inches wide by 2.6 inches high by 12.45 inches deep. The 351 weighs 3.7 pounds and the 350 weighs 3.5 pounds. They cost about $1,400 and $1,100 respectively. The Collins CDIs cost about $400 to $500 more.

Edo-Aire has two VOR NAV receivers—the R-662 (TSO'd) and the R-552 (not TSO'd). These are available with a glideslope receiver as the R-664 (TSO'd) and R-554 (not TSO'd). These units are compatible with their similar

Collins IND 351 VOR/ILS CDI. (Collins photo)

COMs. They have a non-standard width of 3.25 inches, are 2.62 inches high and 11.44 inches deep, and weigh between 2.9 and 3.5 pounds, depending on the model. Prices range from about $650 for the non-TSO'd straight NAV receiver to about $1,400 for the TSO'd version with glideslope. Compatible Edo-Aire CDIs cost between $635 and $810, and weigh 1.9 pounds.

Navigation indicators

Most NAVs come with their own VOR indicator. You can also buy a separate indicator to suit your needs, which may be a left/right needle, a radio magnetic indicator (RMI), an electronic display that simulates needle movement, or simply a digital readout of VOR radials and bearings. Some displays require a *converter* to translate the VOR signal into a needle movement, and others have a converter built in. The standard readout that comes with the NAV set, unless you specify otherwise, will be a course-deviation indicator (CDI) with a nee-

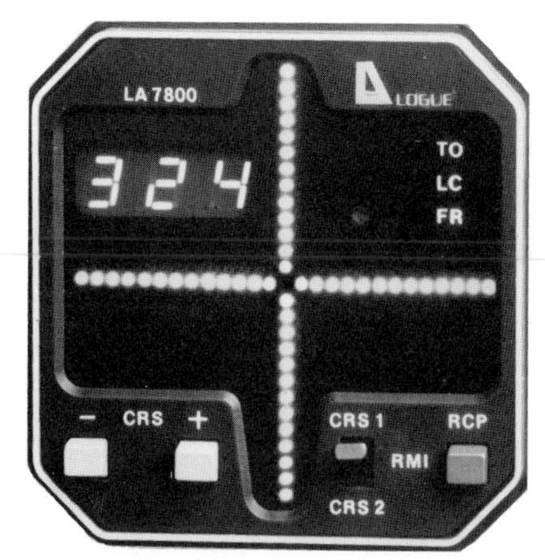

Logue LA 7800 ECDI NAV course indicator. (Logue photo)

The King RMI features one needle for ADF and one for VOR. (King photo)

The HTI VOR/LOC features digital VOR/RMI on one NAV and simultaneous ILS/LOC ECDI on the alternate NAV. (HTI photo)

dle movement. The needle may be only a vertical one to give VOR- and ILS-localizer steering information, or it may have a glideslope needle as well for full ILS. A separate glideslope receiver to use the glideslope needle will be required if there isn't one included with the NAV receiver. Some needles move in rectilinear fashion and others pivot, like windshield wipers. The rectilinear models are preferable, but they are more expensive.

Logue Avionics Ltd., 337 Manchester Road, Poughkeepsie, NY 12603, telephone (914) 471-6210, offers a similar system to the Bendix ECDI, the Digital Nav 7800, which will work with most NAV receivers. As with the Bendix unit, it offers digital radial-readout as well as the ECDI display.

An RMI is similar to an ADF except its compass card rotates to indicate the aircraft heading at all times. Thus needle readings are given as magnetic bearings, rather than relative bearings, as is the case with an ADF. You can get RMIs that will depict both VOR and ADF information with needles. King offers the KI 226 RMI with that kind of display. It requires inputs from a KCS-55 compass system, and costs about $2,200. Collins and Narco offer RMI versions of their ADFs.

A digital RMI is a very inexpensive solution to the the need for immediate VOR-bearing information. Two such units are the Davtron DVOR 902A, priced at about $300 **(Davtron Inc., 427 Hillcrest Way, Redwood City, CA 94062, telephone (415) 369-1188)** and the HTI 100 (under $400)—**Symbolic Displays, Inc., 1762 McGaw Avenue, Irvine CA 92714, telephone (714) 546-0601.**

HTI DVOR/200 digital VOR/RMI. (HTI photo)

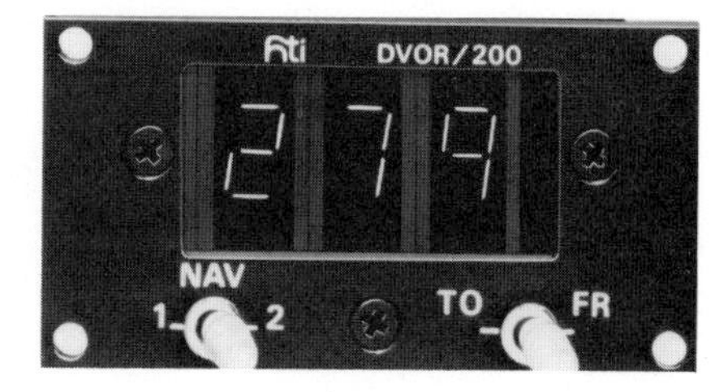

Horizontal situation indicators (HSI)

The HSI, also known as pictorial navigation indicator (PNI), makes navigation easy. It presents a combination of VOR/ILS tracking information and heading data on the same dial, making interpretation much easier. Some HSIs have slaved magnetic compasses as their heading source. Others simply use a directional gyro and must have their heading reset to

The Narco HSI 100 comes in both slaved and unslaved versions. (Narco photo)

conform to a separate compass. HSIs are offered by Bendix (HDS-830—$4,500 and up, slaved only); Collins (PN-101, about $7,000, slaved only); Edo-Aire (NSD-360, slaved or unslaved versions—$2,600 to $4,600); King (KCS-55A, slaved only, about $3,600); and Narco (HSI-100—$2,700 to $3,500, slaved or unslaved).

Distance-measuring equipment (DME)

Distance-measuring equipment (DME) is required for flight above FL 240 (24,000 feet) by FAR 91.33. The DME provides slant-range distance information from VORTACs and TACANs.

The best-selling DME is the TSO'd Narco DME 190/195 series. I have a 190 in my Comanche. The 190 and 195 are basically the same unit, but the 190 is entirely self-contained and mounts in the instrument panel, while the 195 has a small readout instrument that is panel-mounted and a remotely mounted box for all the electronics. The 195 reads out distance, groundspeed, and time-to-station simultaneously. It is remotely tuned by one or both NAV receivers and has a "hold" feature that enables you to stay with the DME you're tuned to even if you retune the controlling NAV receiver to another station. The DME 190 gives the same information, but only one item at a time. It has its own tuner, or it may be remotely tuned. The DME 190 measures 6.25 inches by 2.5 inches by 12.1 inches, weighs 5.2 pounds, and costs about $2,900. The 195 weighs 8.6 pounds and costs about $3,900.

King has two TSO'd DMEs in its Silver Crown line—the KN 62A and the KN 63. The KN 62A is the same size as the KY 196 COM and KN 53 NAV units. It features an electronic readout of either the frequency or the distance, groundspeed, and time-to-station. It can be tuned with its own tuner or from one of your NAV units. It measures 6.25 inches wide by 1.3 inches high by 12.2 inches deep, weighs 2.4 pounds, and costs about $3,100. The KN 63 is essentially a remote version of the KN 62A. It has a small panel-mounted readout of distance, groundspeed, and time-to-station (this may be duplicated on the copilot's side, if desired), and a remote box for the works. It may be channeled from either of two NAV receivers and has the same kind of "hold" feature as the Narco DME 195. It costs about $3,900.

Bendix has a remote DME as

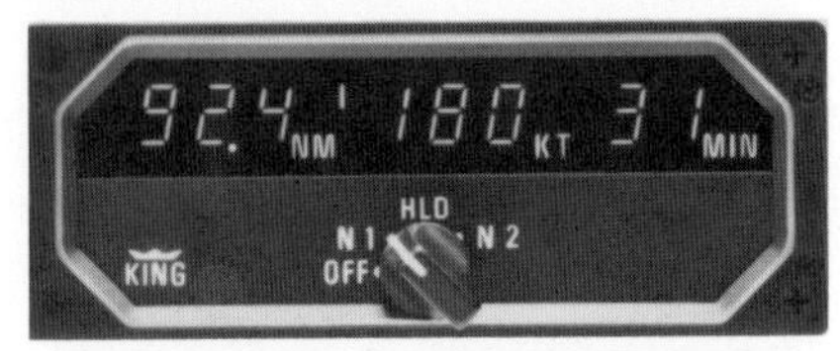

King KN 63 remote DME. (King photo)

part of its BX 2000 system. It features readout of either distance, grondspeed, or time-to-station, and it can also be used as a timer for approaches and flight times. This TSO'd DME 2030 system weighs 8.2 pounds. Its distance information can also read out on the Bendix ECDI. It costs about $4,100.

Edo-Aire offers the RT 888, which is a remote-mounted unit that reads out distance all the time and either groundspeed, distance, or elapsed time. It can be tuned by either of two NAVs, and it has the "hold" feature also found in the Narco DME 195 and King KN 63. The whole system weighs 14.4 pounds and costs about $4,000.

Collins Microline features the DME 451. It is a remote-mounted unit, with a choice of two panel-mounted readouts. The first of these, the IND 451, fits in a standard 3-inch instrument hole and gives distance plus either

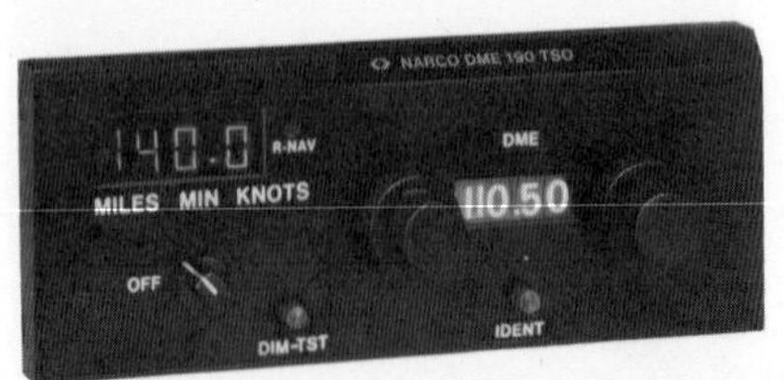

The best-selling Narco DME 190. (Narco photo)

King KN 62 DME. (King photo)

Collins DME 451. (Collins photo)

groundspeed, time-to-station, or elapsed time. The other unit, the IND 450, is very small (3.5 inches wide by 1.35 inches high) and reads each of the three pieces of DME information simultaneously. The whole system weighs 6.4 pounds and costs about $4,000.

Cessna offers the RTA-476A DME in its 400 series. This is a remote-mounted unit weighing a total of 11 pounds. The indicator fits in a 3-inch hole and gives distance plus either groundspeed or time-to-station information. It costs about $3,700.

Cessna RTA 476A DME. (Cessna photo)

Area navigation (RNAV)

Once you have VOR and DME in your aircraft, you have the makings of RNAV. And once you have RNAV you can call the shots in your navigation, because you no longer have to fly from VOR to VOR. With RNAV, you create your own navigation stations, because what you get is a computer that enables you to relocate any VORTAC wherever you want within system limits. So with RNAV you can fly straight lines instead of zig-zags. The relocated VORTAC is sometimes called a "phantom" station, but the correct term is "waypoint." Some RNAVs offer multiple waypoint systems, which can be useful for complex navigation tasks. The important thing to remember is that the RNAV unit you want may not be compatible with existing DMEs and VORs in your panel. Usually, a certain brand will work with like brand units, but even this is not always the case. Make sure that everything works together. You may find yourself having to get a new CDI or HSI to make them work off your new RNAV.

With the aid of the microprocessor, today's RNAVs are incredible pieces of ingenuity, perhaps best exemplified by the King KNS 80, which is an amazing combination of ingredients in one box. What you get for about $5,700 is a VOR receiver, a glideslope receiver, a DME, and a 4-waypoint RNAV. Compare that with some of the DME prices above, and you can see that you get the RNAV for nothing compared to other systems. The whole unit fits in the panel in a 6.25-inch-wide by 3-inch-high by 12-inch-deep space and weighs only 6 pounds!

Foster Airdata, Inc. (7020 Huntley Road, Columbus, OH 43229, telephone (614) 888-9502) offers two RNAV systems, the 511 and 611. The 511 is a two-waypoint system that will work with virtually any VOR or DME. It

Foster Airdata RNAV 511. (Foster Airdata photo)

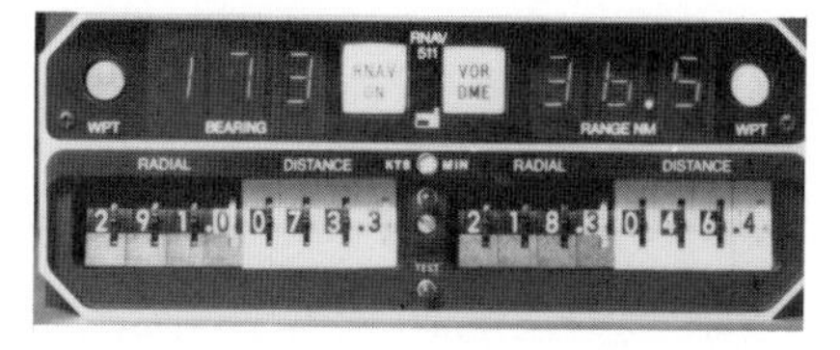

The incredible King KNS 80 RNAV/NAV/DME/GS. (King photo)

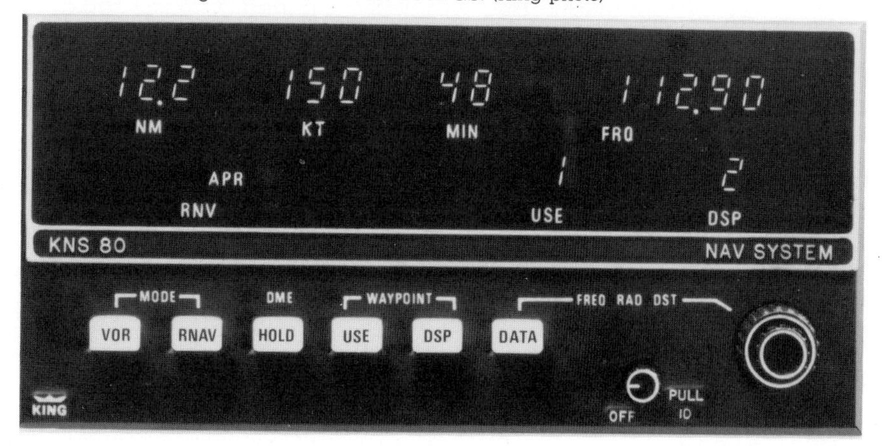

costs between about $1,650 and $2,200, depending on the options ordered, and weighs 2.5 pounds. The 611 offers from 1 to 11 waypoints, and costs between about $4,000 and $6,900. It weighs 10 pounds and is remote mounted.

Collins Microline has the ANS 351 RNAV. This is an 8-waypoint unit that fits in the instrument panel and weighs 3.3 pounds. It costs about $3,100.

Cessna RN 878A RNAV. (Cessna photo)

Collins ANS 351 RNAV. (Collins photo)

Bendix has a unique RNAV system in the BX 2000. The NP 2041A can be preprogrammed with a Texas Instruments SR 52 hand-held calculator, even when you're away from the plane. It has 10-waypoint storage, and can also be programmed from the panel-mounted keyboard/ tuner that is part of the system. It weighs 5.2 pounds and costs about $4,600. The same keyboard that controls the RNAV will also tune frequencies on the associated NAV and COM units in the BX 2000 system.

Cessna offers two RNAV systems, the RN 478A and the RN 878A. The 478A costs about $3,400, while the 878A costs about $5,000. The cheaper version has three waypoints, while the other has five. Both units are compatible with comparable Cessna NAV and DME receivers, but apparently not with competitive makes. The 478A weighs 5 pounds; the 878A weighs 5.1.

Automatic direction finders (ADF)

The ADF is one of the original radio navigation aids and is still a useful device to have on board today, especially in the less heavily populated areas. It operates in the low/medium frequency bands, between 200 and 450 kHz. In addition, ADFs may be used on standard AM broadcasting stations, which means ADF tuners can actually receive between 200 and 1750 kHz. So putting an ADF in your plane not only gives you useful navigation guidance, it also gives you the ability to listen to regular radio broadcasts on those long flights.

In the past all ADFs had analog tuning dials, just like a household radio. However, of late the crystal-controlled ADF has become the industry standard, making tuning a snap instead of the hunt it once was.

Nearly all avionics manufacturers offer ADFs. Perhaps the best feature available on an ADF, "low signal light," is offered by only one firm—Narco. For some reason the other manufacturers haven't seen the need for this. Yet it is essential for safe ADF operation, in my opinion. This is because of the passive nature of the ADF needle, which points to the station. So if you're flying to the station, the needle will sit on the nose of the airplane symbol on the dial—the needle will point straight ahead. If the station goes off the air or the signal is otherwise lost, the needle will continue to point straight ahead, and you could keep flying forever waiting for the needle to swing around, indicating you're over the station. However, with the Narco ADF (TSO'd), a red light illuminates on the dial if the signal is no good. If your unit does not have a low signal light, the only way to prevent confusion is to keep the identification-tone volume up, which means you have to listen to incessant morse-code beeping as you fly.

The Narco Centerline ADF 141 is the standard 6.25 inches wide, 1.75 inches high, and 9 inches

Narco ADF 141 tuner. (Narco photo)

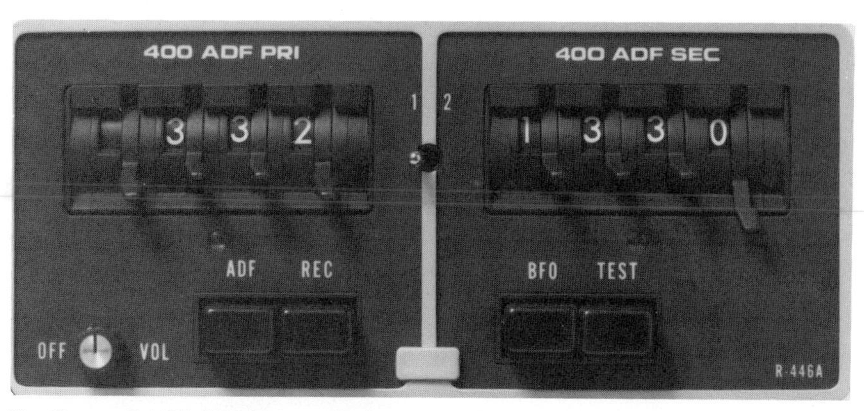

The Cessna R 447A ADF features dual minilever tuning. (Cessna photo)

deep, and weights 6.8 pounds. It costs about $1,700. The ADF willl drive an RMI, if desired.

King now offers the KR 87 ADF, which includes two timers—one to measure flying time and one to time approaches and such. This unit features the gas-discharge frequency display and the "flip-flop" frequency preselect capability to be found in the rest of their Silver Crown line. I have one in my Comanche. It has the same dimensions as the matching KY 196 COM, KN 53 NAV, and KN 62A DME, and costs about $1,900. It offers a combined loop-sense antenna, eliminating the need of some ADFs for the long antenna wire normally strung from the roof to the tail.

King KR 87 ADF. (King photo)

If panel space is at a premium, King offers its KR 86, which puts the ADF dial right in with the tuner. This unit costs about $1,500 and measures 6.25 inches wide by 2.6 inches high by 9.05 inches deep. It weighs 6.6 pounds.

King KR 86 ADF. (King photo)

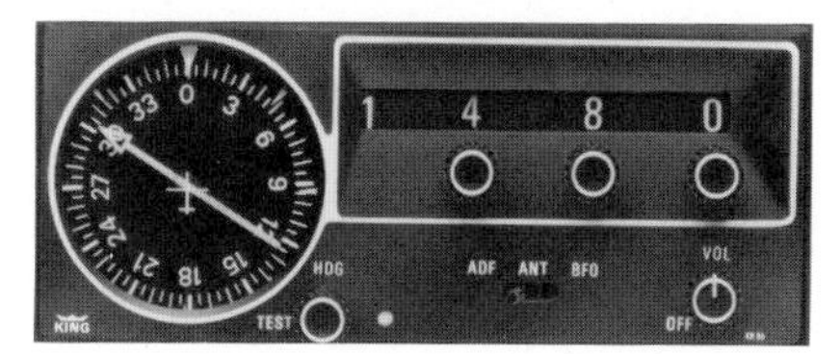

The Cessna model R 446A ADF is tuned by minilevers, a good feature, and offers a dual tuning head, so you can preselect the next frequency. This preselect feature is useful on an ADF, since many IFR approaches require an ADF for the final approach fix; you must retune your ADF for the missed approach fix, if you need it. With frequency preselect, it's simply a matter of pushing a button, rather than retuning at a critical point in the flight. The Cessna R 446A costs about $2,300. Cessna's cheaper ADF has only one tuning head, but it is digitally tuned. This model sells for about $1,600.

Edo-Aire offers an ADF with an integrated azimuth dial and built-in elapsed timer. Although the frequency presentation is electronic and digital, the tuner is analog, not crystal controlled. This unit measures 6.5 inches wide, 3 inches high, 8.75 inches deep, and weighs 8.5 pounds.

The Bendix BX 2000 series ADF 2071A offers an extended-range mode, making it useful for long-distance flights. It is crystal controlled and TSO'd, and measures 6.5 inches wide by 1.75 inches high by 7 inches deep. It weighs 5.5 pounds and costs about $2,000. It offers a combined loop-sense-COM antenna for reduced weight and drag.

Collins offers the ADF 650 in its Microline. This weighs 5.25 pounds, measures 6.25 inches wide by 1.75 inches high by 9 inches deep, and costs about $1,650. It is TSO'd and also offers a combined loop-sense antenna.

Transponders and encoders

The FAA requires a transponder and encoding altimeter for all flight within controlled airspace above 12,500 feet (except that portion within 2,500 feet of the ground), and in certain terminal control areas (TCAs). This is spelled out in FAR 91.24. The transponder *must* be TSO'd. Non-TSO'd transponders were available up to 1975, and so some are still to be found in light aircraft.

Their use is illegal under any circumstances in the United States. Some non-TSO'd transponders can be brought up to the TSO standards by the manufacturer, while others cannot. What do you do with a non-TSO'd transponder you can't TSO? Sell it abroad! Some countries' rules do not, as yet, call for a TSO'd transponder.

A transponder is a device that receives an interrogation signal from an air traffic control (ATC) radar, and sends back a coded reply. The pilot can set the transponder to any of 4,096 codes. If the aircraft has an encoding altimeter, its altitude is sent also. The coding is read by the ATC computer, and generates an alphanumeric data block that appears on the radar screen beside the aircraft blip and moves across the screen with it. The data block reads out the aircraft identification number, its altitude, and its speed. The ATC computer also compares all of the transponder returns it is looking at, and can warn of a possible collision hazard when it predicts that two or more targets will come too close to each other. Normally, the ATC controller assigns each aircraft its own individual code when dealing with it. If an aircraft has not received an assigned transponder code and is flying according to visual flight rules (VFR), it "squawks" the standard VFR code, 1200. Other special squawks denote an emergency (7700), a radio failure (7600), or a hijacking (7500).

The controller can ask any aircraft to "squawk ident." When so instructed, the pilot pushes a button on the transponder, and this causes the appropriate radar target to glow brightly, positively identifying the aircraft.

Narco AT 150 squawking 1200. (Narco photo)

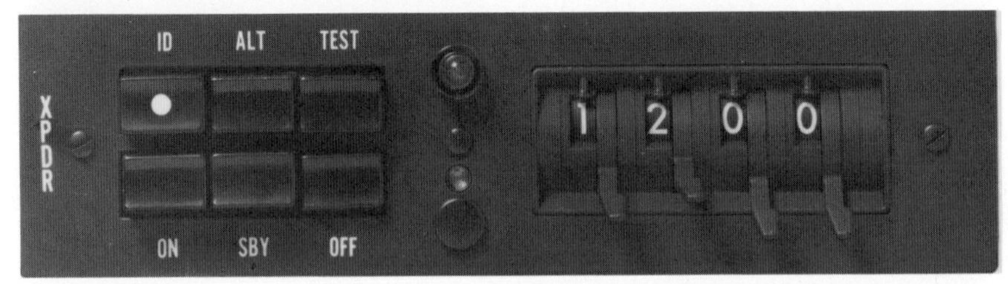

The Cessna RT-859A transponder features minilever controls. (Cessna photo)

Transponders are made by most manufacturers and have few differences. Most of them are "flat-packs" of whatever standard width their manufacturer conforms to. Examples are made by ARC/Cessna (RT-359A/459A/859A), Bendix (TR-2061A), Collins (TDR 950), Edo-Aire (RT-777), Genave (Beta 5000), King (KT-76), Narco (AT 150), and Radair (250). One exception to the flat-pack shape is the Edo-Aire RT-667, which will fit in a standard three-inch instrument hole. This could be useful if you have no space left in your radio stack, but you do have a spare instrument hole. Transponders cost between $600 and $1,200.

Encoders can be built into an existing altimeter, or can stand alone with no visual readout ("blind-encoders"). They must always be connected to a transponder. Various models are available from Bendix, Cessna/ARC, Edo-Aire, King, and Narco. Others are offered by instrument builders, for example:

Cessna 400 series encoding altimeter. (Cessna photo)

Aero Mechanism
7750 Burnet Avenue
Van Nuys, CA 91405
(213) 782-1952

Aerosonic Corporation
3312 Wiley Post Road
Carrolton, TX 75006
(214) 233-8004

Aircraft Instrument and Development
317 East Lewis Street
Wichita, KS 67202
(316) 265-4271

Instruments & Flight Research
2716 George Washington Boulevard
Wichita, KS 67210
(316) 684-5177

Kollsman Instrument Company
Daniel Webster Highway South
Merrimack, NH 03054
(603) 899-2500

Inexpensive encoding altimeters cost between $700 and $1,500. Exotic models can cost even more, some over $8,000.

Audio control panels

If you have a complex avionics installation in your aircraft, it is a good idea to have an audio-control panel. This is a basic switchboard that lets you control which radio you want to hear through your headphones, which through the cabin speaker, which you don't want to hear at all, and which transmitter your mike is on. Many units also include a 75-mHz marker-beacon receiver and its three-light indicator, and an audio amplifier, as well. Units are made by Bendix, Cessna/ARC, Collins, Edo-Aire, King, and Narco. They are all of about the same size and function, and cost about $600.

The King KMA 24 audio control panel allows the pilot to select any combination of radios on either the cabin speaker or headphones. It includes a 3-light 75-mHz marker beacon. (King photo)

Collins AMR-350 audio control panel has toggle switches, giving the pilot a choice of either cabin speaker or headphones for each radio, but not both. (Collins photo)

Radar altimeters

The radar altimeter detects your height above the surface of the earth, using radar signals rather than taking a barometric measurement of the atmosphere. The instrument is important for pilots who fly precision approaches to very low weather minimums. Some are equipped with a selectable "decision-height annunciator," which beeps when you reach a preselected altitude.

The lowest-cost radar altimeter is the Bonzer Minimark, which costs about $1,000 **(Bonzer, Inc., 90th and Cody, Overland Park, KS 66214, telephone (913) 888-6760).** Bonzer makes another relatively low-cost model—the Mark 10X ($2,300)—as does King (the KRA-10, at about $2,200). Everything else costs between $3,000- and $15,000.

King KRA-10 radar altimeter. (King photo)

Weather radar

Weather radar is used primarily to detect thunderstorm activity. It does this by reading radar returns from precipitation. It can also be used for a limited form of ground mapping, particularly for detecting coastlines or rivers. The newest radars are available with color presentation and can also be used to display navigation data, derived from compatible navigation radios in the airplane, and alphanumeric data such as checklists, navigation routes, emergency procedures, and so on.

For many years, radar was not available in single-engine aircraft, because there was nowhere to put the radome and its antenna. This has all changed now, and various systems are available. One uses an antenna in a pod mounted on the leading edge of the wing, or under it, like a drop tank (Bendix), another buries the antenna in the leading edge of the wing (RCA Weatherscout), and another puts the antenna in the nose, beneath the prop, and fires its radar signal through an interrupter device (Robertson Weathersync antenna system for an RCA Primus 20 radar: **Robertson Aircraft Corporation, 839 West Perimeter Road, Renton, WA 98055, telephone (800) 426-7692, or (206) 228-5000 in WA**). This is similar in concept to the mechanism used to fire machine guns through the propeller in World-War I fighter planes.

The Robertson RCA Weatherscout wing-mounted radar antenna on a Cessna 210. (Robertson photo)

Bendix offers color or monochrome radar in all its models, as well as optional checklist and navigation displays. The lowest-cost system is the RDR 160 (about $7,800 in monochrome and $11,250 in color). The checklist is programmed through a hand-held Hewlett-Packard HP-67 calculator.

The Robertson Weathersync antenna goes in the nose of a Bonanza. (Robertson photo)

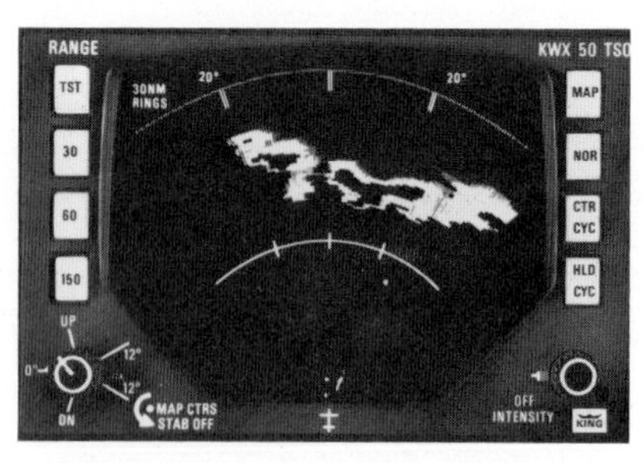

King KWX 50 weather radar, available only in monochrome. (King photo)

Collins does not make a radar suitable for use in single-engine aircraft. Color is available. Collins starts their radar line with the WXR 150 ($7,240) and moves up from there.

King offers its KWX 50 in monochrome only (they're working on a color radar). This unit is currently only available on twins and costs about $6,800.

RCA (**RCA Avionics Systems, 8500 Balboa Boulevard, Van Nuys, CA 91409, telephone (213) 894-8111**) offers its Weatherscout I (about $5,600) for single-engine aircraft and several other models, including some with color.

Ryan stormscope

An alternative to radar is the unique Ryan Stormscope. This is now FAA-approved as a weather-avoidance system where one is required (such as in commuter airlines), and it has been extensively tested by the USAF. The Stormscope does not measure precipitation. It measures lightning flashes, and presents a 360° depiction of these to the pilot on a CRT, which fits in a standard 3-inch instrument hole. It works just as well on the ground as in the air, and will fit any single-engine aircraft. It costs about $6,000 (**Ryan Stormscope, 4800 Evanswood Drive, Columbus, OH 43229, telephone (614) 885-3310**).

Emergency locator transmitters (ELT)

All aircraft, with a few exceptions, must be equipped with a functioning emergency locator transmitter (crash beacon). The exceptions are jets, airliners, single-seater airplanes, crop dusters, certain training aircraft, and aircraft undergoing test flights (see FAR 91.52). ELTs come with their own batteries, and therein lies a snag. Many ELTs were fitted with lithium-dioxide batteries, and these were found to be dangerous, so the FAA prohibited their use. Thus a special rule was enacted permitting aircraft that had been fitted with a lithium-dioxide-powered ELT to fly without an ELT. New standards for lithium-dioxide batteries were then published, and owners were given more time to fit new batteries that met the standards—though no acceptable batteries were available at first.

Many pilots are switching to non-lithium batteries. ELTs that don't use lithium-dioxide batteries are available from **Emergency Beacon Corporation, 15 River Street, New Rochelle, NY 10801, telephone (914) 235-9400,** in several models: the EBC 102A, with 150-mile range ($230); the EBC

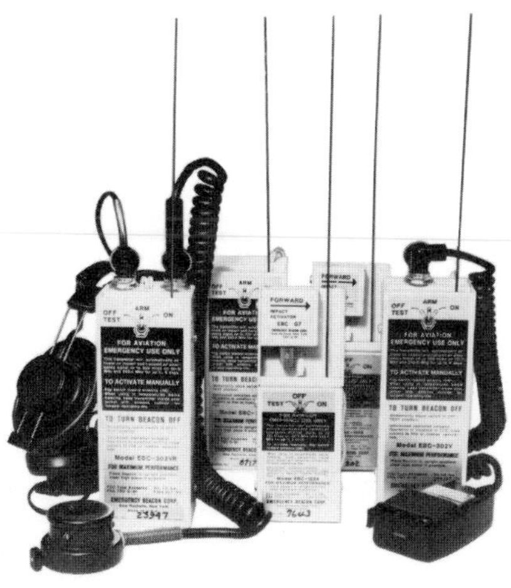

ELTs by Emergency Beacon Corporation. (Emergency Beacon Corp. photo)

202B, with 250-mile range ($350); the EBC-302V, with 300-mile range and voice transmission ($420); and the EBC 302VR, with 300-mile range and voice transmission and reception ($635). Other non-lithium ELTs are offered by Narco (ELT-10—$225) and **Merl, Inc., Box 188, Meriden, CT 06450, telephone (203) 237-8811** (URT-33, starting at $215).

Narco ELT-10 emergency locator transmitter. (Narco photo)

Airborne telephones

Airborne telephone coverage in the United States is not yet complete. There are still large areas of the midwest and west that are not covered. However, coverage in the east and far west is good, and an airborne telephone can be purchased for under $2,000. King sells the KT-96 for $1,255. **Astronautics Corporation of America, Box 523, Milwaukee, WI 53201, telephone (414) 671-5500,** makes models ranging from the SS-II at $595 to the SS-IIID at about $2,800. **Wulfsberg Electronics, 11300 West 89th Street, Overland Park, KS 66214, telephone (913) 492-3000** makes two models, one at about $2,100 (Flitefone II/SE) and one for about $3,300 (Flitefone III).

King KT 96 radiotelephone and handset. (King photo)

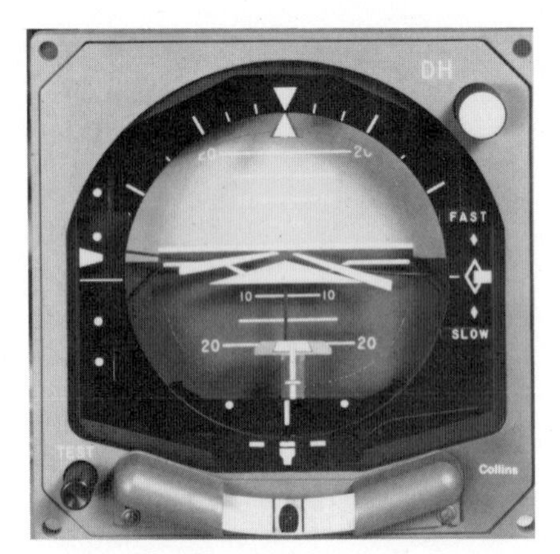

The Collins FD 109 flight director horizon stacks on top of the . . . (Collins photo)

. . . FD 109 HSI, which includes DME readout. (Collins photo)

Flight directors

The flight director takes navigation and altitude informaton and processes it through a computer. Then it gives the pilot commands to fly the aircraft in such a way as to achieve the desired objective. Thus the task of flying the airplane on instruments is vastly simplified. For example, if the pilot wanted to turn right 90°, from 180° to 270°, the flight director heading selector would be rotated to the new heading, and the director horizon would command that the aircraft be put into a right turn. First it would command a bank to the right, and then, when the correct amount of bank has been applied, it would command that this be retained. Finally, when the desired heading is coming up, it would command a roll out of the turn, with the wings to be level when the heading is correct. It will make similar commands for climbs and descents and for capturing and tracking radio navigation courses. Most flight directors are interfaced with an autopilot (in which case they are called integrated flight-control systems), so the commands will in fact be carried out by the autopilot, in which case the pilot is reduced to the task of monitoring the whole thing!

Flight-director systems can cost as much as $30,000. However, there are a couple of lower-cost models. One is the King KFC-200, which costs about $10,500, and Edo-Aire offers its Century IV at about $14,000.

Autopilots

Most autopilots are sold along with the aircraft as an optional extra. The models offered by Cessna/ARC are only available on Cessna aircraft, but various other autopilot models are available for factory installation or retrofit in various existing aircraft. Autopilots can have single-axis control (ailerons only), or two-axis (ailerons and elevators) or three-axis (ailerons, elevators and rudder) controls. The simplest form is the single-axis, which is often called a wing-leveler. Many autopilots also offer navigational couplers. These take radio navigation signals (VOR or ILS) and make the airplane fly along a pilot-selected course. Autopilots can cost from under $1,000 up to almost $10,000. Here are the principal autopilot makers:

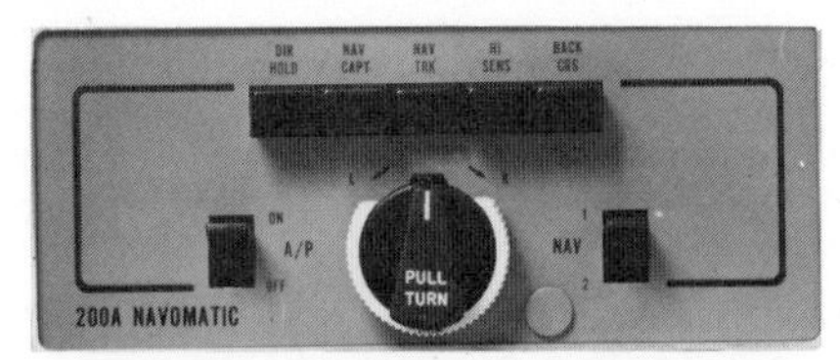

Cessna's low-cost 200A autopilot. (Cessna photo)

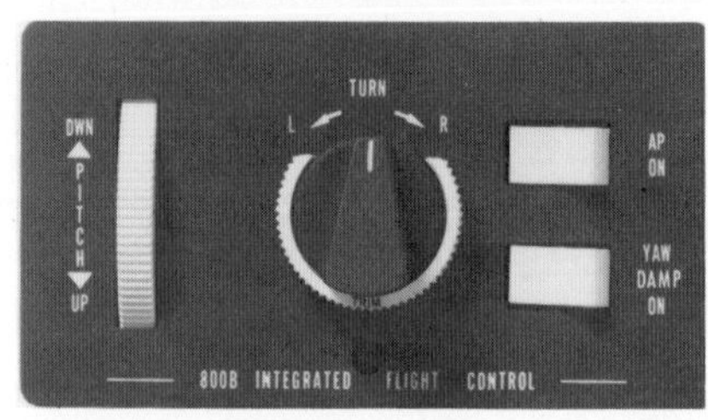

The Cessna 800 series 3-axis autopilot includes an optional yaw-damper. (Cessna photo)

Astronautics Corporation of America
Box 523
Milwaukee, WI 53201
(414) 671-5500

The Brittain B-5 autopilot features heading select, navigational capture and tracking, and altitude hold. (Brittain photo)

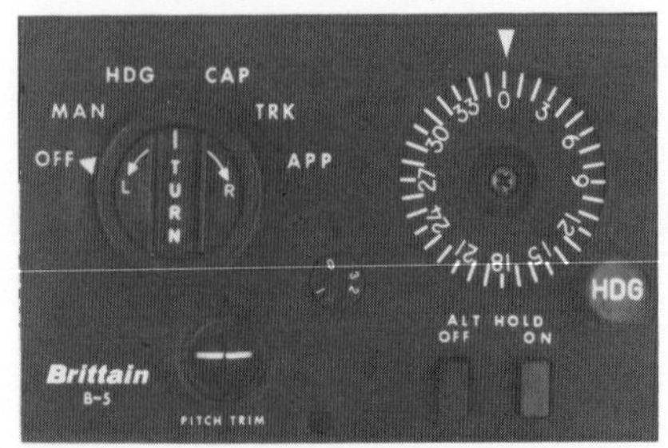

Brittain Industries
Box 51370
Tulsa, OK 74151
(918) 836-7701

Edo-Aire/Mitchell
Box 610
Mineral Wells, TX 76067
(817) 325-2517

Antennas

A good antenna and antenna installation are just as important as good radios. Antennas of all kinds are offered by these makers:

Antenna Specialists
12435 Euclid Avenue
Cleveland, OH 44106
(216) 791-7878

Comant Industries
3021 Airport Avenue
Santa Monica, CA 90405
(213) 390-6694

Communications Components Corporation
3000 Airway Avenue
Costa Mesa, CA 92626
(714) 540-7640

Dayton Aircraft Products
Box 14070
812 NW 1st Street
Fort Lauderdale, FL 33302
(305) 463-3451

Dorne & Margolin, Inc.
2950 Veterans Memorial Highway
Bohemia, NY 11716
(516) 585-4000

Headsets and microphones

There is a little electronic item called *sidetone*. Sidetone is what you have when you hear your own voice through the headphones as you transmit. This makes using your radios much easier, because you can hear yourself talk, and you'll be less likely to shout. I strongly recommend the use of a good headset when flying. Also, for clear transmissions, one of the most important things to have is a noise-cancelling mike. This mutes loud background noise, while taking your voice signal at full strength.

I don't understand why some pilots use hand-held microphones. Your hands should be used to fly the airplane, not to hold a mike! The best thing to use is a headset with a boom mike attached. Some hand-held mikes are so badly designed that you have to look at them first to figure which way you should hold them! Hand-held mikes should be strictly for stand-by purposes. The type of mike that mounts on a gooseneck attached to the cabin wall is a poor alternative.

There is a variety of mikes available for use in aircraft. The most common are the carbon, dynamic, and electret types. The carbon mike is generally standard equipment in most planes, yet it is the worst type to have in an aircraft! Its chief features are low cost and ruggedness. Its chief drawbacks are that it provides only fair frequency response, intelligibility, and noise cancelling capabilities. The dynamic and electret mikes are much better, offering excellent frequency response and intelligibility. They are also more costly. So if you have a carbon mike in your aircraft, relegate it to back-up use when you get your dynamic or electret boom mike!

You can get headphones with heavy sound insulation pads for use in high noise environments,

Telex Communications offers this free booklet (see address below). (Telex photo)

The Narco M700B noise-canceling mike is made by Shure. (Narco photo)

The David Clark headset is one of the best available. (David Clark photo)

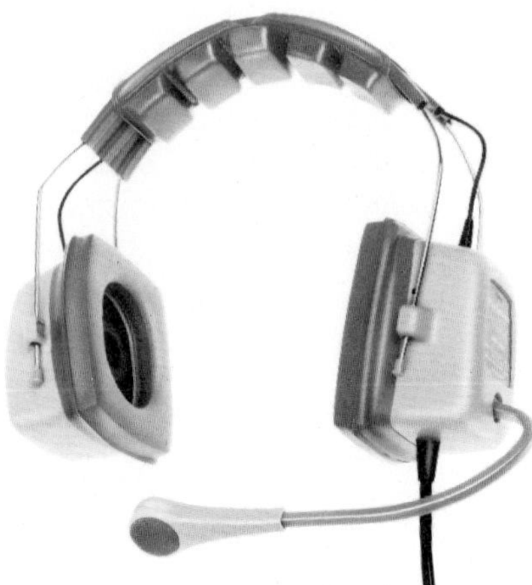

The Telex HearDefender headset comes with either dynamic or electret mikes. (Telex photo)

and you can get very small, lightweight phones that work where the background noise isn't too bad. So the first thing to decide is how much protection against noise you need. Two types of headphones are available—magnetic and dynamic. You need *aviation* headphones. Ordinary stereo phones won't do.

Here are the principal headset/mike manufacturers:

ACS Communications
4865 Scotts Valley Drive
Scotts Valley, CA 95066
(408) 438-3883

David Clark Company, Inc.
375 Franklin Street
Worcester, MA 01604
(617) 756-6216

Electro-Voice, Inc.
600 Cecil Street
Buchanan, MI 49107
(616) 695-6831

Plantronics Inc.
345 Encinal Street
Santa Cruz, CA 95060
(408) 426-5858

Telex Communications, Inc.
9600 Aldrich Avenue South
Minneapolis, MN 55420
(612) 884-4061

Buying new avionics

Most avionics equipment is sold either through the aircraft manufacturer and installed when the aircraft is built, or through a franchised dealer of the avionics maker. Some new avionics, notably those made by Narco, Genave, and Edo-Aire, find their way into mail-order catalogs. However, avionics makers won't honor their warranties unless the radios are installed by one of their dealers, and a dealer won't be too happy about installing a radio you bought at a discount through the mail when he could have sold it to you himself.

Buying used avionics

Used radios are another matter. Many dealers will take your old radios in trade, and some companies have made a business out of the used-radio trade-in market. One, for example, **Connecticut Avionics & Aircraft, Inc. (CAA), Box 555, East Granby, CT 06026, telephone (203) 653-4501** offers to buy virtually any radio a dealer has taken in trade and will later sell it back to the dealer for the same price if he needs it and it's still available. This means a dealer can make an offer on a trade-in he might not otherwise want to take, dump it fast, and still make use of it later if the need arises. CAA maintains a computerized inventory of used radios, and listings are available on request. I have dealt with them myself, and I'm pleased with the results. CAA also has a list of avionics shops that will install used radios bought from them. They offer a 90-day warranty on all used radios.

Other companies that buy and sell used avionics include:

Airwich Avionics, Inc.
1611 South Eisenhower
Wichita, KS 67209
(316) 942-8721

American Avionics, Inc.
7001 Perimeter Road South
Seattle, WA 98108
(206) 763-8530

Avionics Sales Corp.
Box 80220
Atlanta, GA 30341
(404) 455-0348

Memphis Avionics, Inc.
3781 Premier Cove
Memphis, TN 38118
(800) 238-5982
(901) 362-8600

Engines

Aircraft engines are only built by a few manufacturers in the United States—you can count them on the fingers of one hand. Most general aviation aircraft are powered by engines built by Lycoming (**Avco Lycoming, Williamsport Division, 652 Oliver Street, Williamsport, PA 17701, telephone (717) 323-6181**), or Continental (**Teledyne Continental Motors, Aircraft Products Division, Box 90, Mobile, AL 36601, telephone (205) 438-3411**). Jacobs radial engines are now built by Page Industries, Inc., and are available through **Aircraft Engine Service, Inc. (5414 North Rockwell Avenue, Bethany, OK 73008, telephone (405) 789-2711)**. The old Franklin engines were sold lock, stock, and cylinder barrel to Poland a few years ago; they are starting to appear on some European aircraft. The U.S. distributor of Franklin engines is **Carl F. Baker Company (14807 Aetna Street, Van Nuys, CA 91411, telephone (213) 786-3120)**.

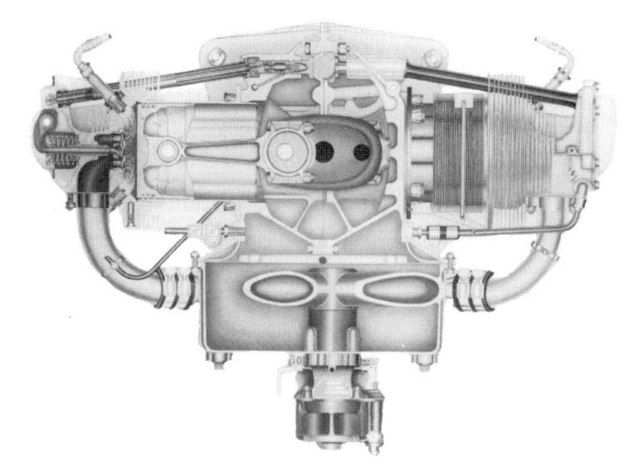

Inside a Lycoming 0 320. (Avco-Lycoming photo)

Lycoming I0-720 400 hp 8-cylinder engine. (Avco-Lycoming photo)

Most light aircraft piston engines are air-cooled, horizontally opposed motors with four, six, or eight cylinders. They have a coding system that describes the make-up of the engine. Each engine has a string of letters and numbers appearing after the manufacturer's name, e.g., Lycoming O-540 A1D5, or Continental GRSIO-520L. The letters indicate the format of the engine, and the first number basically indicates the engine's capacity in cubic inches. The coding system is as follows:

Code	Meaning
A	Aerobatic—limited inverted flight system
AE	Aerobatic—full inverted flight system
G	Geared—the crankshaft rotates faster than the propeller
GS	Geared, supercharged—i.e., not turbosupercharged
H	Horizontal—refers to horizontal mounting of a helicopter engine
I	Injected—fuel injection
L	Left-hand rotation —refers to the counter-rotating engine of a C/R airplane, such as the Piper Navajo C/R
O	Opposed—horizontally opposed cylinders
R	Radial—radially mounted cylinders
T	Turbocharged—Lycoming nomenclature
TS	Turbosupercharged—Continental nomenclature
V	Vertically mounted—refers to helicopter engines

The old small Continental engines used a slightly different description, such as A65-8F or C90-8F. The 65 or 90 refer to the horsepower of the engine. The numbers and letters following the three-digit capacity number refer to the various accessories mounted on the engine. Thus a Lycoming O-540 AlD5 found on a Piper Comanche could not be mounted in place of the Lycoming O-540 J3C5D in a Cessna Skylane RG, although they are basically the same engine, because the accessories won't match. See the table on pages 142–143 for a listing of Continental and Lycoming engines and their characteristics.

Engine TBOs

Aircraft engines are given a suggested lifespan by their manufacturers which is called "TBO"—time between overhauls. Most piston engines have TBOs of between 1,200 and 2,000 hours. Turbine engines, with their simpler construction and smaller number of moving parts, can run as high as 3,000 hours or more before overhaul. I recall that the Rolls-Royce Darts on the Vickers Viscount airliner went up to 7,000 hours between overhauls with no problem. The TBO is a recommended number of operating hours based on the manufacturer's experience with the engine, and is based on proper engine handling and maintenance techniques. A newly designed engine starts with a low TBO, which, as the years go by, may be increased. You don't *have* to comply with the TBO—it's just a recommendation from the manufacturer. Of course, if you exceed your TBO and then have an engine failure, your insurance company will look at you askance.

U.S. Reciprocating Aircraft Engines

■ CONTINENTAL ENGINES

Model	Cyls.	HP	Octane	TBO (hours)
A 65	4	65	80	1800
C 85	4	85	80	1800
C 90	4	90	80	1800
C 125	4	125	80	1800
C 145	4	145	80	1800
E 185	4	185	80	1500
O 200	4	100	80	1800
E 225	4	225	80	1500
O 300	6	145	80	1800
GO 300	6	175	80	1200
IO 360	6	210	100	1200/1500*
TSIO 360	6	200	100	1400
O 470	6	230	80	1500
IO 470	6	260	100	1200/1500*
TSIO 470	6	260	100	1400
IO 520	6	320	100	1200/1500*
TSIO 520	6	300	100	1400
GTSIO 520	6	375	100	1200

■ LYCOMING ENGINES

Model	Cyls.	HP	Octane	TBO (hours)
O 235	4	118	100	2000
O 320 A	4	150	80	1200/2000*
O 320 B/D	4	160	100	1200/2000*
O 320 H	4	160	100	2000
O 360	4	180	100	1200/2000*
IO 360	4	260	100	1200/1600/2000*
TO 360	4	200	100	1200
GO 480	6	260	80	1400
O 540 A	6	250	100	1200/2000*

*TBO varies with models

■ **LYCOMING ENGINES**

Model	Cyls.	HP	Octane	TBO (hours)
IO 540 C	6	250	100	1200/2000*
O 540 E	6	260	100	1200/2000*
IO 540 E	6	290	100	1400
TIO 540 A	6	310	100	1500/1800*
TIO 540 F	6	325	100	1600
TIO 540 J	6	350	100	1600
TIO 541 E	6	380	100	1600
TIGO 541 D	6	450	100	1200
IO 720 A	8	400	100	1800

*TBO varies with models

Engine overhauls

An engine should be overhauled at the time determined by the mechanics and the aircraft owner. The simplest overhaul is called the *top overhaul.* This does not involve a complete teardown, but consists of the replacement or repair of parts outside the crankcase, including the cylinders, pistons, valves, and so on. The *major overhaul* calls for the whole engine to be stripped down to its basic components, and replacement or repair of all the parts to bring them up to approved tolerances. Some engines are completely rebuilt at the factory and are given *zero time*—the equivalent of a virtually new engine. Others are rebuilt by a facility to lower tolerances, and then the engine time is measured *SMOH*—since major overhaul. Sometimes an overhaul involves the installation of chrome cylinders, in which case the time would be measured *SCMOH*—since chrome major overhaul.

The FAA publishes a free Advisory Circular AC 43-11, *Reciprocating Engine Overhaul Terminology and Standards,* obtainable from DOT Publications Section, M-443.1, Washington, DC 20590. **Aviation Consumer (1111 East Putnam Avenue, Riverside, CT 06878, telephone (203) 637-5900)** publishes a booklet *Anatomy of an Engine Overhaul,* obtainable for $6.45 including postage.

Some shops that specialize in engine overhauls are:

AAR of Oklahoma, Inc.
Box 19508
Oklahoma City, OK 73144
(405) 681-2361

Air Engines, Inc.
P.O. Drawer T
Sanford, FL 32771
(800) 327-9432
(305) 323-3831 in FL

City Aviation, Inc.
Box 221
Northampton, MA 01060
(413) 584-1860

G & N Aircraft, Inc.
1701 East Main Street
Griffith, IN 46319
(800) 348-4061
(800) 552-8956 in IN
(219) 924-7110

Mattituck Airbase, Inc.
Mattituck, NY 11952
(516) 298-8330

Midwest Aviation Service, Inc.
Route 5
Ottumwa, IA 52501
(515) 682-8733

Page Industries, Inc.
Box 191
Yukon, OK 73009
(405) 354-5385

Pryor Aircraft Engines of Tennessee, Inc.
420 Railroad Street
Elizabethton, TN 37643
(615) 542-2811

Schneck Aviation, Inc.
Box 6417
Rockford, IL 61125
(815) 965-4001

Schneck Aviation, Inc.
SW Division
1222 99th Avenue
San Antonio, TX 78214
(512) 924-9261

T. W. Smith Engine Company, Inc.
Hangar 1
Lunken Airport
Cincinnati, OH 45226
(513) 871-3500

Norm Bender in Memphis, TN, sells *brand-new* Lycoming engines on an exchange basis for less than the price the factory charges to rebuild one. This is because he sells your old engine on the open market, where he can get a much better price than Lycoming gives on an exchange for either a rebuilt or a new engine. He requires the old engine to be in airworthy condition—to the extent that you have to fly in to his base and complete a normal shutdown procedure. He offered these figures in late 1979:

Engine (all Lycomings)	HP	Factory new price	Factory rebuilt exchange	Norm Bender new price exchange
O-360-A1A	180	$8,420	$6,364	$5,795
O-540-A1A5	250	$12,024	$9,068	$7,995
IO-540-N1A5	260	$13,992	$10,552	$9,995
IO-720-A1B	400	$24,040	$18,116	$16,995

Norm Bender points out that these figures are based on trade-ins of first-run engines—those that have never been overhauled, or those that are no more than 100 hours past the recommended TBO. Otherwise there is an extra charge of $400. Installation is extra. He stresses that these prices are subject to change. **Norm Bender Incorporated, Box 30343, Memphis International Airport, Memphis, TN 38130, telephone (901) 365-6611.**

Turbochargers

Some engines can accept turbocharging, which is a system that takes the exhaust gas and runs it through a turbine to drive a compressor. The high-density air from the compressor goes into the induction system and provides greatly enhanced performance at high altitudes. **Rajay Industries, Inc. (Box 207, Long Beach, CA 90801, telephone (213) 426-0346)** offers turbochargers for many aircraft.

Further references

Avco Lycoming publishes a free newsletter called *Avco Lycoming Flyer*, which contains many useful hints on operating their engines. Ask for it by writing to them at **Williamsport, PA 17701.** Ask for a copy of *Key Reprints*, as well, which will give you important back issues. The same people also have a handy booklet *Information Guide for Avco Lycoming "Direct Drive Engine Operation,"* which is based on a slide presentation of the same name. It can be obtained from the Training Department at the above address.

Teledyne Continental has a booklet, *Pilot's Handbook On Engine Operation*. This is available for 50¢ from them at **Box 90, Mobile, AL 36601.**

Fuel

Light aircraft are among the most energy-efficient means of transportation available. Consider the table of seat-miles-per gallon figures given on pages 62–63. Here we see numbers like 76.2 seat miles to the gallon for a Beech Bonanza A 36, compared to 43.4 for a Boeing 747; 63.2 for a Piper Archer II compared to 42.8 for a Lockheed 1011; 50.4 for a Cessna 402C compared to 41.9 for a McDonnell Douglas DC 10.

Aviation gasoline today comes in 80- and 100-octane grades. The 80-octane grade is sometimes hard to find, and as a result some people have to use 100 octane in an engine not designed for it. The fuel companies have developed a low-lead-content fuel—100 LL—to help in this area, but there are still many reports of valves burning out when this fuel is used in an 80-octane engine. You can tell what grade of fuel you have by its color, as follows:

Red: 80/87 octane (1/2 ml of lead per gallon)

Green: 100/130 octane (3 to 4 ml of lead per gallon)

Blue: 100/130 octane low lead (2 ml of lead per gallon)

Clear: aviation jet fuel

The octane rating denotes the antiknock qualities of the fuel. The higher the rating, the more compression the fuel can stand without detonating. Detonation occurs when the fuel-air mixture inside the cylinder burns too rapidly after the spark plug fires. Engines are designed so that the mixture will burn at a certain controlled rate. The more compression the fuel can stand without detonation, the more power can be developed from it. The first of the two numbers in the fuel octane designation (e.g., 100/130) indicates the lean-mixture rating (as in cruising flight), and the second the rich-mixture rating (as in takeoff or climb).

Lead is put in the fuel to increase its octane rating. However, in a lower compression engine, designed to run on 80-octane fuel, high lead content

When Beech brought out the Turbo A36TC Bonanza, they proudly boasted it in big letters on the tail—until some lineboys thought that it meant it took jet fuel! Jet fuel doesn't do a piston engine any good at all. (Beech photo)

can increase lead deposits in the engine's combustion chamber. This can lead to valve sticking, detonation, preignition, and all sorts of nasty things. If you have to use 100-octane fuel in an 80-octane engine, you should avoid running at a mixture that is too lean.

TCP (tricresyl phosphate) has been approved by the FAA as an additive in non-supercharged Continental and Lycoming engines. One quart of TCP treats about 300 gallons of fuel. TCP is available from Alcor, Inc., Box 32516, San Antonio, TX 78284, telephone (512) 349-3771.

Fuel suppliers

Here are the principal aviation fuel suppliers in the United States:

Amoco Oil Company
200 East Randolph Drive
Chicago, IL 60601
(312) 856-5111

Chevron International Oil Company
114 Sansome Street
San Francisco, CA 94104
(415) 894-7800

Continental Oil Company
Box 2197
Houston, TX 77001
(713) 965-1534

Exxon Company, USA
Box 2180
Houston, TX 77001
(713) 656-3636

Gulf Oil Company
Box 1519
Houston, TX 77001
(713) 750-2000

Mobil Oil Corporation
150 East 42nd Street
New York, NY 10017
(212) 883-4041

Phillips Petroleum Company
308 Adams Building
Bartlesville, OK 74004
(918) 661-4507

Shell Oil Company
Box 2105
Houston, TX 77001
(713) 241-5524

The Standard Oil Company (Ohio)
Midland Building
Cleveland, OH 44115
(216) 575-4141

Texaco Inc.
2000 Westchester Avenue
White Plains, NY 10650
(914) 253-4000

Union Oil Company of California
1065 East Golf Road
Schaumburg, IL 60196
(312) 885-5293

Oil

Just as an aircraft engine won't run without fuel, it won't run without lubricating oil for too long. Oil not only lubricates the engine, it also helps to cool it and clean it of deposits. There are two basic types of oil used in aircraft—straight mineral oil and ashless dispersant (AD). Most engines are broken in with straight mineral oil and then are switched to the AD-type after the first 100 hours or so.

There are three weights of aviation oil, which are classified under an SAE (Society of Automotive Engineers) system. The three grades of oil refer to their viscosity. The viscosity of oil is its resistance to flow. You need low viscosity oil in cold temperatures, and high viscosity oil in warm ones. Here are the grades:

Commercial Aviation #	Commercial SAE #	Viscosity
65	30	Low
80	40	Middle
100	50	High

The oil can will have these numbers embossed on the lid; the letter W, if it appears, indicates the oil is the ashless-dispersant type.

Oil analysis

It is a good idea to have your engine oil analyzed after each oil change. This is done by a variety of companies, using an emission spectrometer. By flash burning the oil, they can determine its metal and other contents, thus giving you an idea of the amount and type of engine wear. Oil analysis should be carried out on an ongoing basis, so you can determine a trend. A single analysis won't be too helpful.

Oil analysis can detect the following elements:

Substance	Wear Item
Aluminum	Pistons, bearings
Chromium	Rings, cylinders
Copper	Bearings
Iron	Rings, crankshaft, camshaft
Lead	(Gasoline additive)
Magnesium	Piston rings
Nickel	Piston rings, bearings
Silicon	Dirt (through air intake)
Silver	Bearings
Tin	Bearings

The elements are given in parts-per-million (PPM) by weight in the oil. Some labs also report the oil's viscosity, acidity, and sludge content.

Here are some companies that provide oil analysis:

Ana-Lab
111 Harding Avenue
Bellmawr, NJ 08030
(609) 931-0011

Analysts, Inc.
Box 7111
Oakland, CA 94601
(415) 536-5914

Analysts, Inc.
Box 226
Linden, NJ 08036
(201) 925-9393

Lymco Laboratories
Suite 300
6400 West Park Plaza
Houston, TX 77057
(713) 783-9140

Spectra-Check
13600 Deise Avenue
Cleveland, OH 44110
(216) 451-6455

Spectro, Inc.
Box 16526
Fort Worth, TX 76133
(817) 292-2646

Wear-Check
87 Walton Street
Atlanta, GA 30303
(404) 522-7385

Microlon

The FAA recently approved a "miracle additive" called Microlon, which is claimed to improve an engine's wearability. It does so by coating the walls of the cylinders with a minute, smooth layer of Teflon. Most users have reported improved performance and reduced oil consumption. Microlon is available from **Chemlon, Inc., 4055 Hollister, Houston, TX 77080, telephone (713) 462-5553.**

Modifications

There is a whole industry that specializes in modifications to aircraft after they are built and sold by the original manufacturer. Some modifications are made many years after the aircraft was produced, and are designed to bring an aging beauty up to modern standards. Others are carried out on new aircraft to improve their performance characteristics.

Typical modifications include:

Performance improvements

- Turbocharging
- Fuel-injection
- More powerful engines
- STOL kits
- Aerodynamic clean-ups
- Twin-engine from single-engine

Utility improvements

- Increased fuel capacity
- Camera holes
- Larger freight doors
- Landing-gear conversions
- Metalizing fabric covering
- Pressurization
- Extra seats

Aesthetic improvements

- One-piece windshields
- New instrument panels
- Extra side windows
- New wing tips

Modifications to an FAA type-certified aircraft must be made under a **Supplemental Type Certificate** (STC). An STC is expensive to obtain, taking many man-hours to comply with FAA requirements. This is why you will usually see firms specializing in STCs for one type of plane, or a similar modification to a variety of aircraft.

For example, **Robertson Aircraft Corporation, 839 West Perimeter Road, Renton, WA 98055, telephone (800) 426-7692, (206) 228-5000 in WA and outside the continental United States,** specializes in STOL (short-takeoff and landing) conversions of many popular light aircraft. They offer kits on these models:

- Beech Bonanza V 35 A & B
- Cessna 150
- Cessna 172 Skyhawk
- Cessna 180
- Cessna 182 Skylane
- Cessna 182 Skylane RG
- Cessna 185
- Cessna 205
- Cessna 206 Stationair 6
- Cessna 207 Stationair 7
- Cessna 210 Centurion
- Cessna 310
- Cessna 336 Skymaster
- Cessna 337 Super Skymaster
- Cessna 340
- Cessna 401
- Cessna 402
- Cessna 414
- Cessna 421 Golden Eagle
- Piper Aztec
- Piper Comanche
- Piper Cherokee
- Piper Cherokee Six
- Piper Twin Comanche
- Piper Seneca

Robertson also holds STCs to put a wing-mounted radar antenna on the Centurion and nose mounted antenna on the Bonanza.

Part of the Robertson STOL conversion on a Cherokee Six. (Robertson photo)

Cessna 310s show off their Robertson STOL modifications. (Robertson photo)

Avcon Industries, Inc., Box 4248, Wichita, KS 67204, telephone (316) 838-9375, produces powerplant upgrades for the Cessna 150, 170, 172, 175, 177, Piper Cherokee 140, 150, 151 and 160.

B & M Aviation, 2048 Airport Way, Bellingham, WA 98225, telephone (206) 676-1750, offers a flap and aileron gap-seal kit for most single-engine Cessnas. It is supposed to improve performance.

Beryl D'Shannon Aviation Specialties, Route 2, Box 272, Jordan, MN 55352, telephone (612) 492-2611, offers all kinds of improvements for the Beech Bonanza including more power, better windows, wingtip fuel tanks, modernized instrument panels and so on. They also offer a one-piece windshield for the Piper Comanche. I have one on my Comanche.

Ralph Bolen, Inc., 1311 US 40 SW, London, OH 43140, telephone (614) 852-1990, provides taildragger conversions for the Cessna 150, 152 and 172.

The Colemill President 600-Baron mod. (Colemill photo)

Bush Conversions, Box 431, Udall, KS 67146, telephone (316) 782-3851, offers STOL kits on most single-engine Cessnas, the Cessna Skymaster and the Piper Cherokees, as well as more power for the Cessna 150, 170, 172, 175, and the 180 hp Cardinal.

Colemill Enterprises, Box 60627, Nashville, TN, telephone (615) 226-4256, produces 300 hp engine mods for the Beech 55 Baron (the Baron "President 600") and the Cessna 310 (the Three Ten "Executive 600"). They also do a number on the Piper Navajo, calling it the "Panther Navajo," which incorporates a 350 hp engine conversion, four-bladed props and other improvements.

Custom Aircraft Conversions, 234 West Turbo Drive, San Antonio, TX 78216, (512) 349-6347, offers a tailwheel conversion for the Cessna 150-152—the "Texas Taildragger."

Flint Aero, 336 Front Street, El Cajon, CA 92020, telephone (714) 448-1551, produces auxiliary fuel tanks for almost all the single engine Cessnas. Some of these mount internally in the wing, while others add a couple of feet to the span.

The Colemill Executive 600. (Colemill photo)

Horton Aero, Inc., Wellington, KS 67152, telephone (800) 835-2051, (316) 326-2241 in KS and outside the continental Unmited States), offers STOL kits for most single-engine Cessnas, the Skymaster and Piper Cherokees.

Light Plane Components, 509 Buckley Road, Aurora, CO 80011, telephone (303) 366-5897, offers a tailwheel conversion for the Piper Tri Pacer and Colt. They also have a larger engine available.

Madras Air Service, Box 1225, Route 2, Madras, OR 97741, telephone (503) 475- 2360, offers "super tips"—huge sweeping wingtips for most Beech Cessna and Piper single-engine aircraft, as well as the Stearman and Stinson 108.

Met-Co-Aire, Box 2216, Fullerton, CA 92633, telephone (714) 870-4610, offers Hoerner wing-tips for most single-engine Cessnas, Piper Cherokees, Comanches and Twin Comanches, Beech 18s and Bonanzas.

J. W. Miller Aviation, 447 Sandau Road, San Antonio, TX 78216, telephone (512) 349-6820 makes a conversion of the Piper Twin Comanche called the Miller 200. This includes a one-piece windshield, an extended nose with a baggage compartment, engine nacelle baggage compartments and a 200 hp engine conversion (up from 160 hp), among other things. The same company offers radomes for older Aero Commander twins.

Night Sign Inc., Suite 2310, 400 Mansion House Center, St Louis, MO 63102,. telephone (314) 421-5466, offers an electronic aerial billboard that mounts on the bottom of many aircraft and helicopters for aerial advertising use.

The Night Sign aerial billboard. (Night Sign photo)

Pearce Aeronautics, Inc., 120 North Old Manor Road, Wichita, KS 67208, telephone (316) 685-4552, has increased power mods for the Cessna 172 and the Piper Cherokee 140 and 150.

RAM Aircraft Modifications, Inc., Box 4580, Waco, TX 76705, telephone (817) 799-0264, produces engine and other modifications for the Cessnas 172, 206, 310, 340, 414 and the Piper Cherokee 140 and 150s.

Riley Aircraft Corporation, 2386 Palomar Airport Road, Carlsbad, CA 92008, telephone (714) 438-0660, improves the power on the Cessna 310, 340 and 414, offers a turboprop mod for the Cessna 421.

Seguin Aviation, Box 225, Seguin TX 78115, telephone (512) 379-3278, does a very nice job on the Old Piper Apache, producing the *Geronimo*, which has a considerable aerodynamic cleanup, new 180 hp engines, a long nose, a new tail and more.

Smith Speed Conversions, Inc., Box 601, Johnson, KS 67855, telephone (316) 492-6254, produces aerodynamic improvements to the Beech Bonanza and Baron for extra speed.

Hangars

Hangar space is at a premium at most airports. One solution to this problem is for a group of aircraft owners to get together and build a hangar. In this way, the airport owners still own the land, while the aircraft owners own the hangar space as a cooperative. One developer that has tried this with considerable success is **Lawson Construction, Inc., Box 19268, Portland, OR 97219, telephone (503) 640-3825.** Lawson developed a group of tee hangars at the Portland-Hillsboro Airport, and sold 36 of the first 45 hangars

Shade hangars at Easton, PA. (Tim Foster photo)

The Aerodome hangar. This one's at Wichita, Kansas. (Aerodome photo)

Inside the Aerodome is a turntable. (Aerodome photo)

within the first six months, according to *Airport Services Management* magazine.

Hangars can run from large, exotic buildings, with clear spans and office space adjacent, down to simple tee hangars. Sometimes you'll even see shade hangars—nothing more than a big roof on poles designed to protect aircraft from the hot sun. I thought shade hangars were confined to the warm southern climes until I saw some at Braden's Flying Service in Easton, Pa.

One unique approach to storing more planes in less room is the Aerodome, produced by **Aerodome Industries, Inc., 2716 George Washington Boulevard, Wichita, KS 67210, telephone (316) 684-6266.** This hangar features a turntable floor, so six single-engine aircraft can occupy the space that would normally be taken by two or three tee hangars. Aerodome claims that up to 84 aircraft can be stored in space that would accommodate only 40 aircraft in tee hangars.

Large prefabricated hangars are available from firms such as these:

A & S Building Systems
Box 40099
Houston, TX 77040
(713) 424-2644

Atlantic Building Systems, Inc.
Box 82000
Atlanta, GA 30366
(404) 447-9010

Republic Buildings Corporation
Box 6778
Cleveland, OH 44101
(216) 574-7100

Strat-O-Span Buildings and Materials Company
Breese, IL 62230
(800) 642-9310 in IL
(800) 851-4550 outside of IL

K-Span American Company
Box D
Springdale, AR 72764
(501) 756-6031

Tee hangars or larger buildings can be obtained from local builders or various prefabricated manufacturers, such as:

Erect-A-Tube, Inc.
Box 409
Harvard, IL 60033
(815) 943-4091

Fleetwood Metals, Inc.
355 West Alondra Boulevard
Gardena, CA 90247
(213) 327-1300

Fulfab, Inc.
Box 1365
Canton, Ohio 44708
(216) 477-7211

PVP Enterprises, Inc.
Box 1864
Grand Prairie, TX 75051
(214) 263-2659

Steel Structures of Colorado, Inc.
4388 South Windermere
Englewood, CO 80110
(303) 761-3985

Steel Structures, Inc.
Box 555
Meadville, PA 16335
(814) 336-3159

Umbaugh Builders, Inc.
Box 71
Ravenna, OH 44266
(216) 296-9989

Portable hangars

Some tee hangars are sold as portable units, so you can simply rent some space at the airport and put up your own hangar. **State Recreation, Inc.** (address below) has a convincing argument for the portable hangar concept. It goes like this: airport land leases are usually confined to twenty years maximum at city and county airports. Real estate improvements—for example, permanent hangars—become airport property at the end of the lease. Many FBOs are unable to become strong enough financially to buy hangars, or the remaining term of their lease doesn't justify the expenditure. It is very difficult to get a lease extended before it expires, further aggravating the improvement situation. Airports are often subject to long-term planning studies, which preclude interim construction. Some airports don't allow privately owned portable hangars. Meanwhile the aircraft owner/taxpayer must tie down his or her aircraft where it is exposed to the elements. In California, Proposition 13 increased the tax on personal property vs real estate property, and a portable, privately owned hangar is regarded as personal property. This should give civic leaders there, where there is an acute shortage of hangar space, the incentive to approve the use of portable hangars. Such hangars are offered by:

A portable tee hangar by Port-A-Port. (Port-A-Port photo)

State Recreation, Inc.
3383 East Gage Avenue
Huntington Park, CA 90255
(213) 583-0901

Port-A-Port
Box 1738
Paso Robles, CA 93446
(805) 238-4003
(800) 235-4141 outside CA

Hangar doors

Although most hangars come with doors, you can buy and install customized doors in your own hangar. For example, one type employs a fabric roll-up design, which works like a blind. This lets in the light and keeps the heat in. One brand is called the Para-port and it's sold by:

Kuss Corporation
1331 Broad Avenue
Findlay, OH 45840
(419) 423-9040

Both roll-up and bifold doors are made by Erect-A-Tube (see address above). Bifold doors are offered by:

Mosher Doors, Inc.
Box 309
Riceville, IA 50466
(515) 985-2007

Industrial Door, Inc.
Walworth, WI 53184
(414) 275-6869.

Winches and tows

If your airplane is heavy like mine, and you have a little hill to

The Gettleman Lug-Bug. (R. McCoy photo)

climb every time you put the bird in its nest, you'll need something more than muscles to move it. You can move it with a winch. Erect-A-Tube (see address above) offers the *Airplane Winch* with a 1/2-hp electric motor for planes of under 3,500 lbs and a 3/4-hp motor for planes over that. Another winch is the Ekon model, sold by **Thern, Inc., Box 347, Winona, MN 55987, telephone (507) 454-2996.** Or you may prefer a powered tow bar, such as the Lug-Bug, offered by **Gettleman Manufacturing, Inc., Route 3, Box 104, Mineral Wells, TX 76067, telephone (817) 325-3313;** or the Power Tow, which is powered by an electric or gasoline motor, sold by **Phoenix Aviation, Route 3, Box 293, Sandpoint, ID 83864, telephone (208) 263-6632.**

Aircraft and windshield covers

If you can't get your aircraft into a hangar, at least protect it as much as you can. External covers help keep out heat and moisture. Internal heat shields help keep the cabin and avionics cool. Covers are made by:

Airtex Products, Inc.
Lower Morrisville Road
Fallsington, PA 19054
(215) 295-4115

Jon Polcik Company
10949 Tuxford Street
Sun Valley, CA 91352
(213) 768-3684

Pro-Tec-Prop Company
Box 1551
Big Bear Lake, CA 92315
(714) 866-5642.

Internal heat shields are offered by:

Morgan Stanford Aviation
(Thermacon Heat Screen)
2510 Russell Street
Berkeley, CA 94705
(415) 841-6642

Connecticut Aviation Products, Inc.
(Thermoguard Heat Shields)
Box 153, Bissel Station
South Windsor, CT 06074

Preheaters

If you must leave your airplane outside, you may need a preheater for the engine. These are offered by:

Aircraft Components, Inc.
700 North Shore Drive
Benton Harbor, MI 49022
(800) 253-7261
(616) 925-8861 in Michigan

Flame Engineering
(Red Dragon Pre Heater)
Box 577
La Crosse, KS 67548
(913) 222-2873

Wag-Aero, Inc.
Box 181, North Road
Lyons, WI 53148
(414) 763-9588

You can buy Herman Nelson, Janitrol, and American Airfilter heaters through the **Tobias Company, Inc., Box 748, Aurora, CO 80010, telephone (303) 344-0185.** If you have access to electricity, you could install an engine block heater. I have one in my Comanche. It keeps the engine warm as toast, and makes it very easy to start in the winter. Block heaters can be ordered for all Lycoming engines and for Continental O-470, IO-520, and TSIO-360 engines from **Tanis Aircraft Services, Box 117, Glenwood, MN 56334, telephone (612) 634-4772.**

Interiors

It's surprising sometimes to see a beautifully repainted aircraft with a disastrous interior. This doesn't have to be. One company that has been making do-it-yourself interiors for years is **Airtex Products, Inc., Lower Morrisville Road, Fallsington, PA 19054, telephone (215) 295-4115.** I have one of their interiors in my 16-year-old Comanche, and it looks very nice. Airtex has patterns for most light aircraft, and they'll make an interior to order for you to put in yourself. If they don't have the pattern, they can make it up if you send them your old interior, just like a Hong Kong tailor. Most aircraft interiors consist of a headliner, wall panels, seats, and carpeting. Each of these components can be ordered separately, as needed.

If you don't feel like installing an interior yourself, you'll find that many aircraft paint shops also do interiors, or you might consider an automobile upholstery shop. However, bear in mind that Federal Aviation Regulations have specific requirements on the fire resistance of interior fabrics (FAR 23.853).

Redoing your own interior is permitted under the aircraft-owner preventive maintenance rules laid out in FAR 43 Appendix A, provided that no primary structure or operating system must be disassembled. If it *must*, you should have a licensed mechanic approve or supervise the work before returning the aircraft to service.

Paint

Refinishing an old airplane is a good way to reduce its apparent age fast—the aeronautical equivalent of a facelift. Refinishing it is such a complex procedure that it is best left to a professional aircraft paint shop. Improperly applied paint can actually adversely affect the flight characteristics of an aircraft.

There are basically four types of paint available for aircraft use—polyurethanes, modified polyurethanes, enamels, and lacquers. The beautiful wet-look finish comes from a polyurethane paint job. In addition to the high-gloss finish, this paint gives a harder surface and lower maintenance, compared to enamels and lacquers. However, it doesn't "breathe," so corrosion can spread under the paint without being seen. The other paints will crack or otherwise allow such corrosion to show through. Acrylic lacquer is probably the cheapest paint to apply, since it is easy to spray on and it dries quickly, which is why you'll see

An Airtex interior. (Airtex photo)

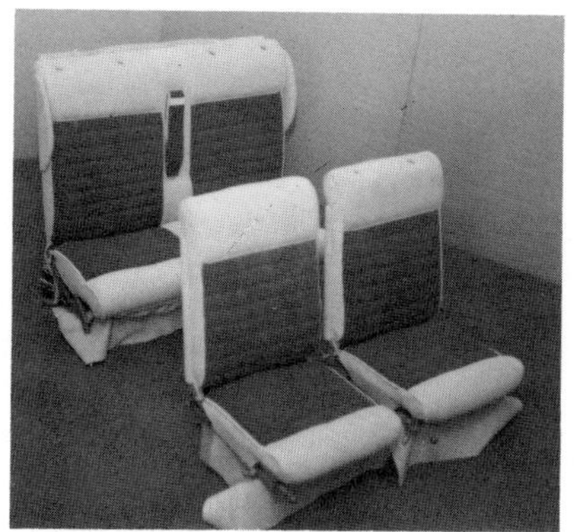

Part of the stripping-down process prior to repainting at Palmer Paint in Wellsville, NY. (Palmer Paint photo)

some of the manufacturers using it. Acrylic enamel gives a good gloss, but not the true wet-look. Most paint shops only use the polyurethane wet-look paints.

To refinish an aircraft properly, it is first stripped down with chemicals, and then the surface is cleaned, polished, and inspected. Next the surface is thoroughly washed and rinsed. Any aluminum surfaces are etched with an acid solution and rewashed. Primers are then applied, after which the basic coat is applied. Any masking is done next and secondary colors are sprayed on. Finally the masking is removed, and there is your paint job, or respray as the English say.

One of the prettiest Ercoupes around is N24AP, painted by Green of Turner Field, Ambler, PA. (Mary Foster photo)

Paint manufacturers

The following firms offer paint especially for aircraft use:

Alumigrip brand
US Paint, Lacquer and Chemical Company
2102 Singleton Street
St Louis, MO 63103
(314) 621-0525

Ameron Industrial Coatings Division
Box 2153
Wichita, KS 67201
(316) 733-1361

Ditzler Aircraft Finishes
PPG Industries, Inc.
Box 5090
Southfield, MI 48037
(313) 275-5550

Imron brand
E. I. DuPont de Nemours Company, Inc.
1007 Market Street
Wilmington, DE 19898
(302) 774-2421

Ranthane brand
Randolph Products Company
Box 67
Carlstadt, NJ 07072
(201) 438-3700

Stits Aircraft Coatings
Box 3084
Riverside, CA 92519
(714) 684-4280.

Complaints

Everyone wants to complain sometimes. Here are some effective ways to complain about aeronautical matters.

Rule 1—go to the top

The best person to complain to is the chairman of the board of the corporation involved. Throughout *The Aviator's Catalog* you will find the addresses and phone numbers of aviation suppliers. So if you are peeved about somthing, write to the chairman at the address given. It's unlikely that the chairman will personally write or call back to you, but he or she can lean more effectively on the offending party than anyone else in their organization, and things generally happen faster when the big boss is calling the shots.

Rule 2—put it in writing

Write a letter that is as bereft of emotional overkill as you can make it (a little sarcasm is permissible, if appropriate). However, your objective is to obtain satisfaction, and since duelling is now out, the best way you can achieve that is to have things put the way they should be. So lay out the facts in a logical and clear manner, with times, dates, people's names, and other references, such as serial or model numbers. Explain what happened and what you are now looking for. Don't be a mean bastard. Save that for the lawyers. Be polite and explicit. Indicate a time frame in which you'd like the problem resolved. Give a phone number where you can be reached. If you haven't heard back within *one week*, send a follow-up letter, with a photocopy of the original. Chances are good that things will by now start to happen.

Rule 3—keep good records

Keep good records of all the correspondence and action taken—what you did, who you saw, phone calls made, and so on. You'll need them if you have to sue.

Rule 4—If at first you don't succeed, stir things up!

If you don't receive satisfaction, let your fellow consumers know by sending copies of the correspondence and an explanatory covering letter to any or all of the following, as appropriate:

Director
Bureau of Consumer Protection
Federal Trade Commission
Washington, DC 20580
(202) 523-3727

The Administrator
Federal Aviation Agency
800 Independence Avenue SW
Washington, DC 20591
(202) 426-4000

The Aviation Consumer
1111 East Putnam Avenue
Riverside, CT 06878
(203) 637-5900

General Aviation Manufacturers Association
Suite 517
1025 Connecticut Avenue NW
Washington, DC 20036
(202) 296-8848

National Business Aircraft Association
One Farragut Square South
Washington, DC 20006
(202) 783-9000

Pilot's Lobby
Box 1515
Washington, DC 20013
(202) 546-5150

If you think others may be able to help, look in the "Associations" section, page 229.

Alas, the beautiful Rockwell Commander 112 TCA is no longer in production. (Rockwell photo)

P A R T F O U R

Aviator's Equipment & Services

AVIATION CATALOGS
DIRECTORIES
CHARTS
WEATHER

Robertson STOL Cessna 421 in flight test.
(Robertson photo)

Aviation Catalogs

Many aviation materials are sold by mail, and to sell these we have the aviation catalogs. One of the best known is "The Flyer" put out by **Aircraft Components, Inc., 700 North Shore Drive, Benton Harbor, MI 49022, telephone (800) 253-7261, (616) 925-8861 from Alaska, Hawaii and Michigan.**

Aircraft Components, Inc. sells sunglasses . . . (Ray-Ban photo)

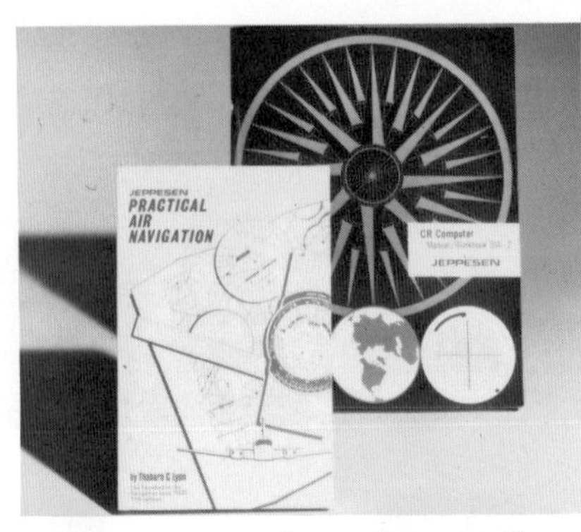

. . . training manuals . . . (Jeppesen-Sanderson photo)

. . . and instruments, among other things. (Aerosonic photo)

This is mailed to about 250,000 people every couple of months. It contains listings of a wide variety of pilot ads, aircraft radios, instruments, gadgets, and parts.

Another big catalog is the one published by **Wag-Aero, Inc., Box 181, North Road, Lyons, WI 53148, telephone (414) 763-9588.** This goes to about 275,000 people and has listings for many aircraft parts and even some homebuilt aircraft designs, as well as the usual array of pilot aids. Wag-Aero is located on its own airstrip (WI92), with 3100-feet E-W and

This tent converts your high-wing plane to a motel. It's called a *Wing-A-Bago*, and comes from Wag-Aero. (Wag-Aero/Raettig photo)

2100-feet N-S turf runways, and welcomes visitors by air or car. They are 1/4 mile north of Highway 36 at 1216 North Road in Lyons.

Sporty's Pilot Shop (Clermont County Airport, Batavia, OH 45103, telephone (513) 732-2411) sends catalogs to about 100,000 people. Sporty's is an actual store that you can go to and shop in. It is located at an airport near Cincinnati. It specializes in the vast array of aeronautical paraphernalia that appeals to many pilots—sunglasses, radios, navigation computers, gadgets, kneeboards, books, watches, headphones, aviation jewelry, and some airplane parts and instruments.

Sporty's creates its own training tapes. (Sporty's photo)

Training records and tapes for the pilot by Jeppesen, via mail-order catalog. (Jeppesen-Sanderson photo)

Practice instrument flying under hazy conditions with these shot glasses from Sporty's. (Sporty's photo)

Pan American Navigation Service offers books, charts, pilot aids, and so on in their catalog. They also have a store, located at 16934 Saticoy Street in Van Nuys, CA. Their mailing address is **Pan American Navigation Service, Box 9046, Van Nuys, CA 91604, telephone (213) 345-2744.**

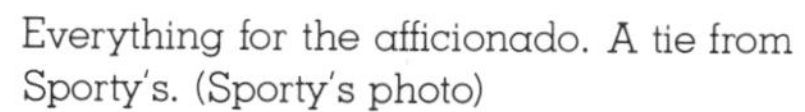

Everything for the afficionado. A tie from Sporty's. (Sporty's photo)

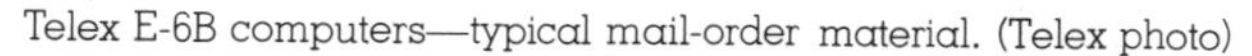

Telex E-6B computers—typical mail-order material. (Telex photo)

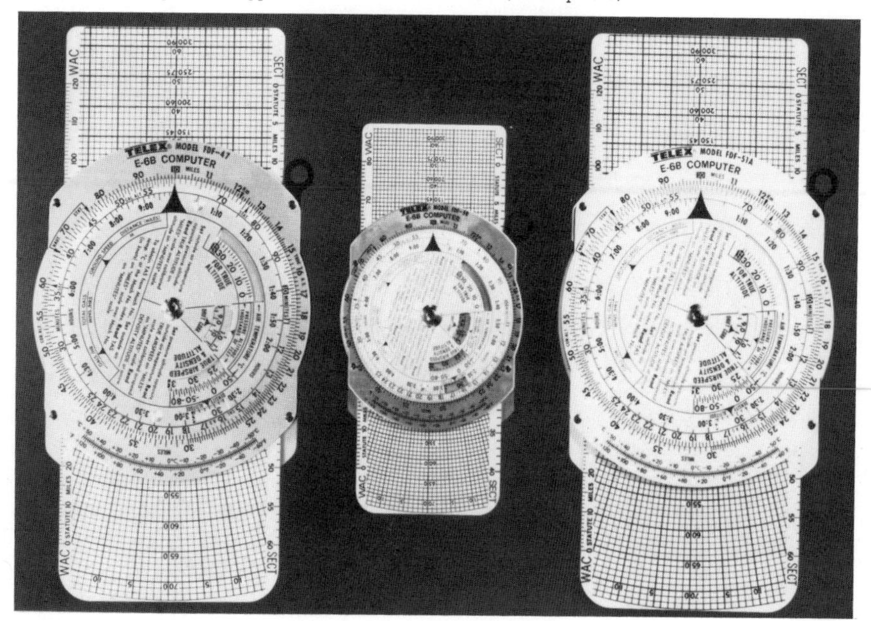

For the more technical aircraft components—e.g., an Ercoupe cabin heat shroud assembly, alternators, engine mounts for various aircraft, and so on—there is the catalog of **Univair Aircraft Corporation (Route 3, Box 59, Aurora, CO 80011, telephone (303) 364-7661)**, which costs $2.00. Univair holds the FAA-type certificates for many out-of-production aircraft, such as Stinsons, Swifts, etc., and actually devotes much of its time to making new parts for old airplanes. They are much more than just a mail-order catalog house.

Many people buy parts through the mail. This Whelen wingtip strobe light is an example. (Whelen photo)

One of the most interesting catalogs is that of **Aircraft Spruce and Specialty (Box 424, Fullerton, CA 92632, telephone (714) 870-7551)**, which costs $3. This has extensive descriptions of the materials for sale that will help you decide on their applicability. This catalog is particularly appealing to the aircraft homebuilder.

Other catalog sources include (free unless noted):

Aircraft instruments

Century Instrument Corporation
440 Southeast Boulevard
Wichita, KS 57201
(800) 835-3344
(316) 683-7571 in KS

Aircraft parts

All Aircraft Parts
16673 Roscoe Boulevard
Sepulveda, CA 91343
(213) 894-9115

Aviation Products
114 Bryant
Ojai, CA 93023
(805) 646-6042

Avmat 20
Box 30325
Memphis, TN 38130
(901) 345-6668, or
(800) 238-6816

B & F Aircraft Supply
6141 West 95th Street
Oaklawn, IL 60454
(312) 422-3220

Cooper Aviation Supply Co.
2149 East Pratt Boulevard
Elk Grove Village, IL 60007
(312) 364-2600

J & M Aircraft Supply Inc.
Box 7586
Shreveport, LA 71107
(318) 222-5749 (price $1)

Aircraft sculptures

Van Guilder
Box 1092
Tustin, CA 92680
(714) 544-6681

Belt buckles

Parkview Distributors
Route 1
Stoddard, WI 54658
(608) 788-0966

Books

Aero Publishers
329 West Aviation Road
Fallbrook, CA 92028
(714) 728-8456

Historic Aviation
3850 Coronation Road
Eagan, MN 55122
(612) 454-2493

Sky Books International
48 East 50th Street
New York, NY 10022
(212) 688-5086

Zenith Aviation Books
Route 2, Box 341
North Branch, MN 55056
(612) 583- 2573

Clothing

Avirex Ltd.
468 Park Avenue South
New York, NY 10016

Flight Apparel Industries
Box 166
Hammonton, NJ 08037
(609) 561-9200

International Military Supply Co., Inc.
9165 Roosevelt Boulevard
Philadelphia, PA 19114
(215) 677-7200

Helmets

Split-S Aviation
529 C Forman Drive
Campbell, CA 95008
(408) 377-1884

Jewelry

V-M Enterprises
19 Maple Drive
Belleville, MI 48111
(313) 697-4963

Parachutes

Midwest Parachute
22799 Heslip Drive
Novi, MI 48050
(313) 349-2105

Paraphernalia

Avirex Ltd.
468 Park Avenue South
New York, NY 10016
(212) 697-3414

Great American Propeller Company
555 Westmont Drive No. 212
San Luis Obispo, CA 93401
(805) 481-4450

Split-S Aviation
529 C Forman Drive
Campbell, CA 95008
(408) 377-1884

This authentic replica of the RAF fighter-pilot's jacket from the Battle of Britain, the Irvin, comes from Avirex. (Avirex photo)

Another Avirex offering, the G-1 goatskin flight jacket, with helmet and goggles. (Avirex photo)

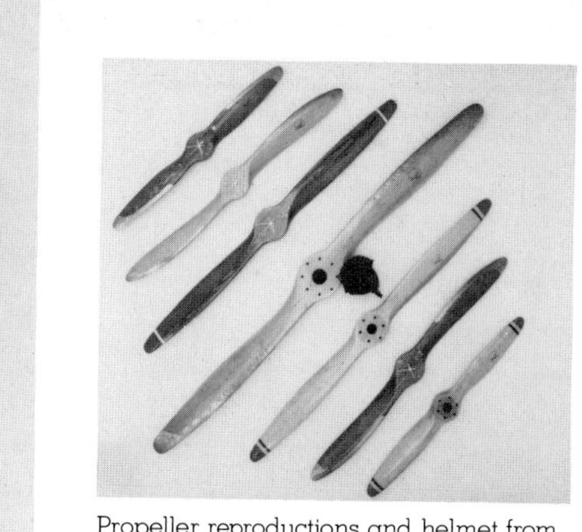

Propeller reproductions and helmet from Split S. (Split S/Horizon photo)

Warbird shotglasses—another Split S offering. (Split S photo)

Pilot Aids

Gladco Aviation
Box 586
Elmira, NY 14902
(607) 562-3578

Vanmark Pilot Center
2801 East 78 Street
Minneapolis, MN 55420
(612) 854-8776

Tools

Aircraft Tool Supply Company
Box 148
Oscoda, MI 48750
(800) 248-0130
(517) 739-1447 in MI

David Andrews Supplies
534 West Avenue
Norwalk, CT 06850
(203) 866-1919

US Industrial Tool and Supply Company
13543 Auburn
Detroit, MI 48223
(800) 521-7394
(313) 272-4545 in MI

Most mail-order houses take credit cards and ship by UPS. Turnaround time is generally quite fast, but sometimes the item won't be held in stock, meaning you may have to wait a while. The Federal Trade Commission has a rule that says you have the right to cancel an order if it is not shipped within 30 days. If you cancel the order, the catalog house has to refund your money forthwith. Using a credit card helps, here, too.

If you're not satisfied with the service you get, you can complain to the **Director, Bureau of Consumer Protection, Federal Trade Commission, Washington, DC 20580.** They wield a big stick.

Directories

There are other sources of information that can be very helpful. One is the *Aviation Buyer's Digest*, **Room 302, One Bank Street, Stamford, CT 06901, telephone (203) 325-2647.** This is published four times a year. It is free to qualified users. Another is the *World Aviation Directory*, published by **Ziff-Davis Publishing Company, 1156 15th Street NW, Washington, DC 20005, telephone (202) 293-3400.** This comes out twice a year and costs $50 per copy. Z-D also publishes *Flying Annual and Buyer's Guide* at the beginning of each year (order from them at **One Park Avenue, New York, NY 10016, telephone (212) 725-3500**), which gives a good look at the current array of aircraft, avionics, and equipment, and the *Planning and Purchasing Handbook* of *Business/Commercial Aviation* in April of each year (order from **B/CA** at **Hangar C-1, Westchester County Airport, White Plains, NY 10604, telephone (914) 948-1912).**

Charts

Aviation charts are used for pilotage by visual or radio navigation according to visual flight rules (VFR), for flight by radio navigation according to instrument flight rules (IFR), and for IFR approaches and departures at airports. They are available from government and commercial sources. All U.S. Government charts are published at regular intervals by the National Ocean Survey (NOS) of the National Oceanic and Atmospheric Administration (NOAA) of the U.S. Department of Commerce. All their charts can be bought by subscription (so that you will always get the newest chart when it comes out) from **NOS, Riverdale, MD 20840, telephone (301) 436-6993,** or through **AOPA, Box 5800, Washington, DC 20014, telephone (301) 654-0500.**

This flight bag will hold 4 Jeppesen manuals, plus. (Sporty's photo)

Visual charts

NOS visual charts come in several formats. The two most commonly used are the Sectional and VFR Terminal Area Charts. Sectionals are produced at a scale of 1:500,000—about 8 miles to the inch. Thirty-seven separate charts are issued, covering the conterminous United States and southern Canada, and are revised and reissued twice a year. Nineteen more charts cover Alaska, Hawaii, and Puerto Rico. Some of these are revised twice a year and some only once a year.

VFR Terminal Area Charts are produced at a scale of 1:250,000—about 4 miles to the

Sectional and VFR Terminal Area Charts

■ **CONTERMINOUS U.S. AND HAWAIIAN ISLANDS**

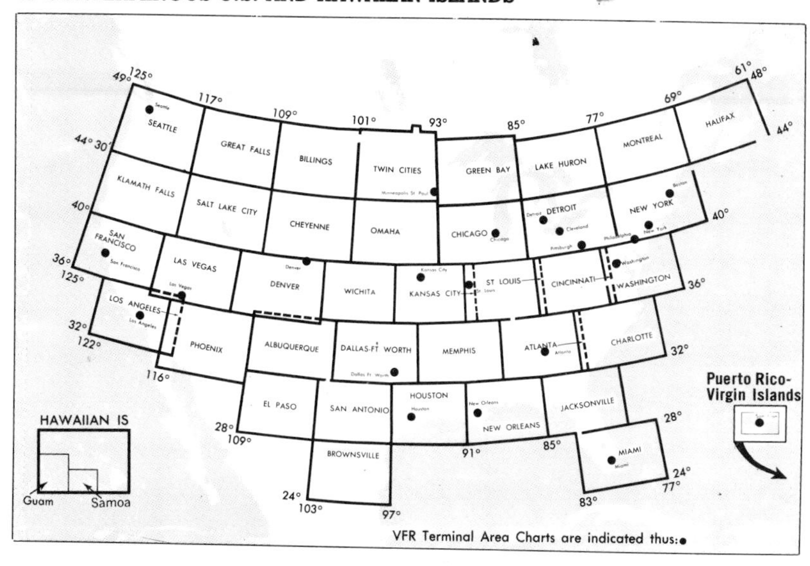

■ **ALASKA**

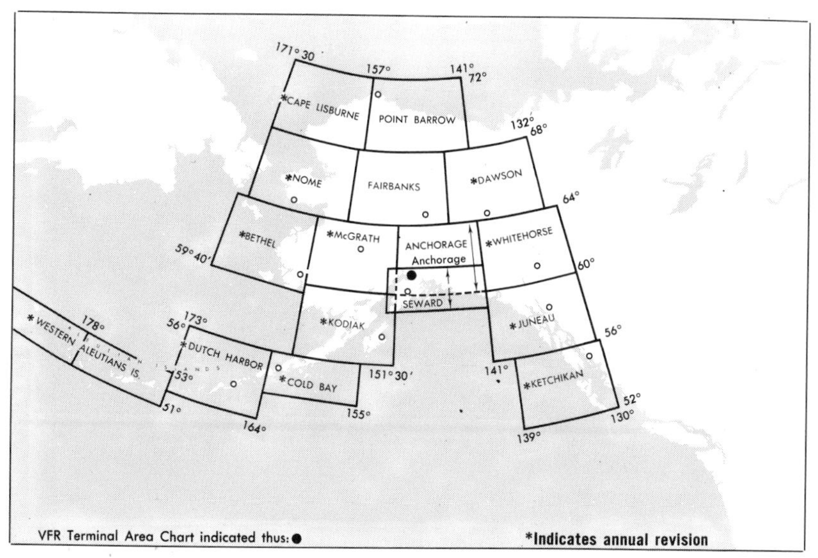

inch. Twenty-one of these are published for most major metropolitan areas, and they are revised twice a year.

Sectional and Terminal Area Charts are available over the counter at most airport dealers, or by subscription, as mentioned above.

A new series of VFR visual/radio navigational charts, called Flightcharts, has been introduced. These come in subscriptions covering six broad areas of the United States, and are produced to a variety of scales, like radio-facility charts (this is an important difference from Sectionals, which have a constant scale). They are printed in a uniform 23" × 19" size, and are reissued every six months. The main benefit of Flightcharts is that the VOR navigational information is easier to read than on a Sectional, and they contain more airport information. Order these from **National Aeronautical Institute, 306 Dartmouth Street, Boston, MA 02116, telephone (617) 247-0600.**

Sky Prints is an annual atlas of VFR radio navigation charts, with direct VOR-to-VOR courses laid in when there is no airway. A monthly update sticker is included in the annual subscription. **Sky Prints Corporation, 6617 Clayton Road, St Louis, MO 63117, telephone (314) 862-2255.**

Every ace has a kneeboard to hold his charts. This one is from Sporty's. (Sporty's photo)

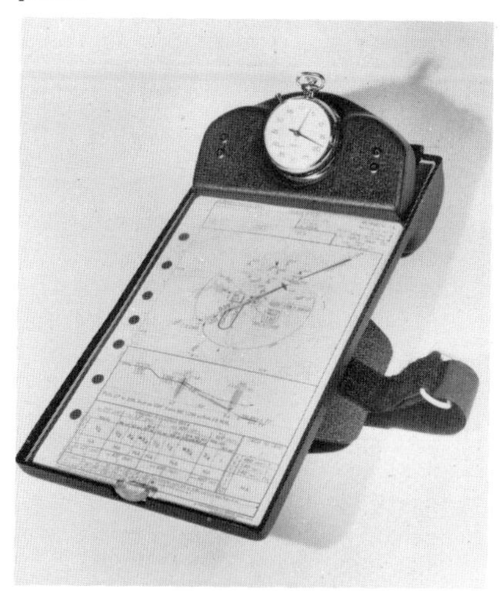

Planning charts

NOS produces a VFR/IFR Planning Chart to a scale of 32 nm to the inch. This comes in two sheets covering the entire United States. One side is VFR and the other is IFR, so you can take your choice on how you use it. It measures 82" × 56" and is suitable as a wall chart. NOS also has a VFR Flight Case Planning Chart, which covers the United States at a scale of 60 nm to the inch. A Gulf of Mexico/Caribbean Planning Chart is published, too.

Other visual charts

Smaller scale charts, called World Aeronautical Charts (WACs) or Operational Navigation Charts (ONCs) are produced at a sale of 1:1,000,000—about 16 miles to the inch. Twelve of these cover the conterminous United States and eight more cover Alaska. Many more cover the rest of the world. They are suitable for VFR navigation by fast aircraft. Jet Navigation Charts (JNCs) come in 1:2,000,000 scale (about 32 miles to the inch) and are available for the whole world. Global Navigation Charts (GNCs) come in 1:5,000,000 scale (about 80 miles to the inch), again with world-wide coverage.

Many other charts are offered through NOS, which publishes a free *Catalog of Aeronautical Charts and Publications*, giving detailed descriptions of these.

IFR charts

NOS publishes a complete set of IFR enroute and approach charts. The approach charts are issued in bound books and reissued every 56 days. Both low- and high-altitude charts are available, covering both sets of airways. These may be ordered from NOS or AOPA on a subscription basis, as indicated above. You can also buy charts

World Aeronautical and Operational Navigation Charts

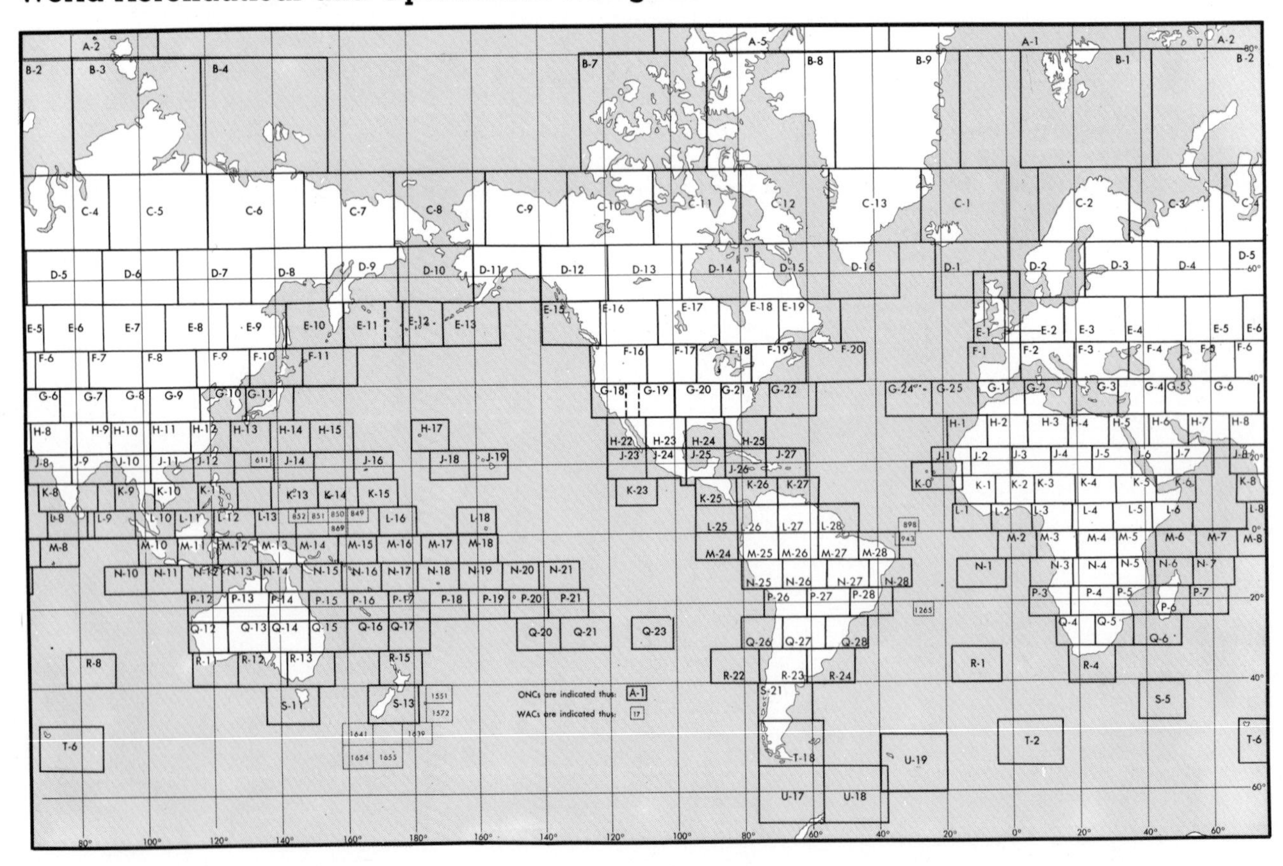

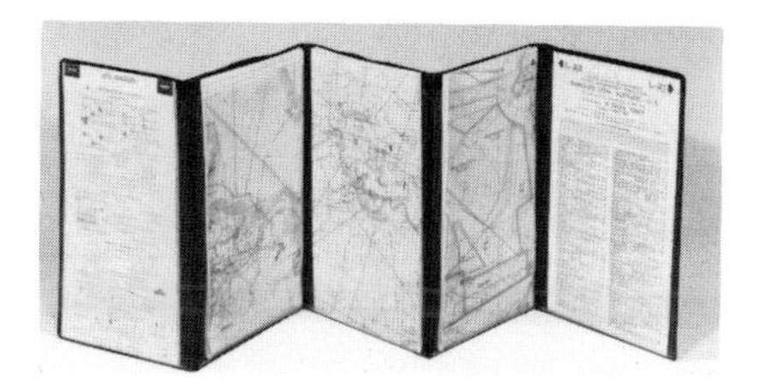

This chart case holds NOS visual and IFR charts. From Sporty's Pilot Shop. (Sporty's photo)

Enroute Low Altitude Charts

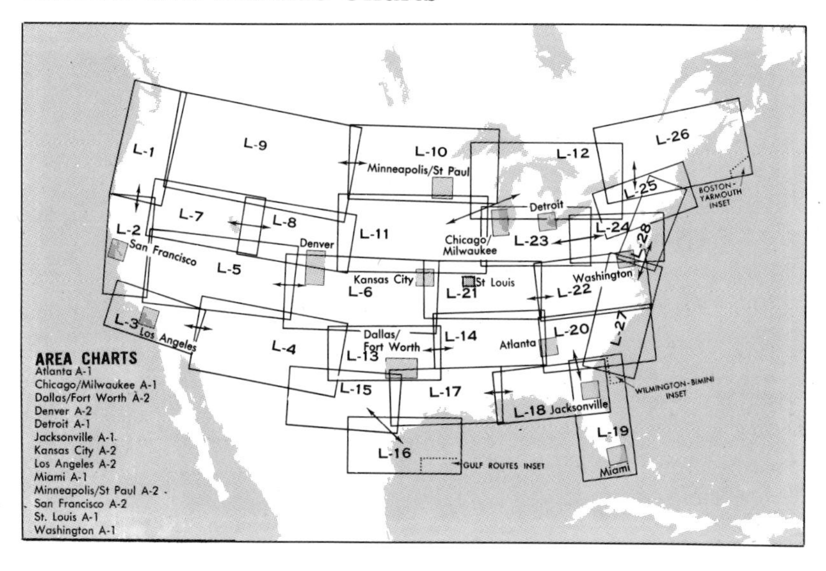

Instrument Approach Procedure Charts

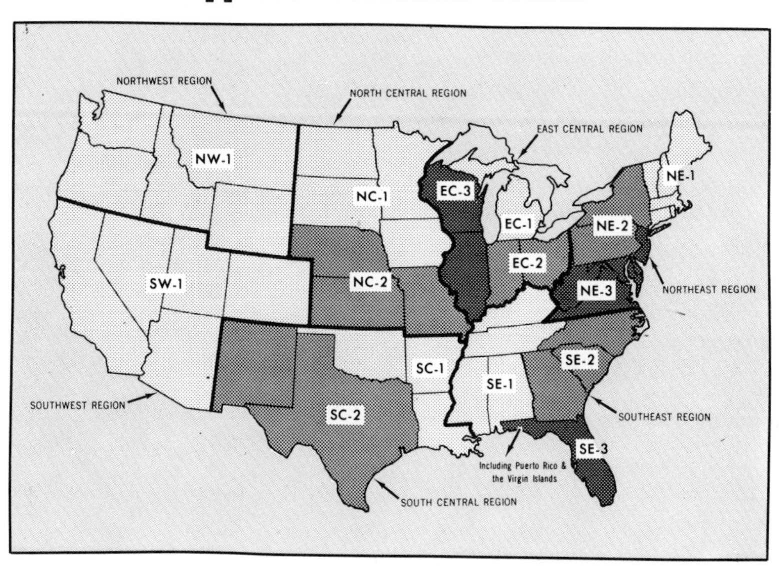

as a one-shot deal, which is useful for special trips.

Jeppesen publishes its own set of charts, and provides worldwide coverage on a customized subscription basis. Jepp charts are issued in loose-leaf binders and are revised every week. Jeppesen also issues a VFR/IFR RNAV (area navigation) chart series. Jepp charts may be ordered through **Jeppesen Sanderson, Box 3279, Englewood, CO 80112, telephone (303) 779-5757.** If you are a Jeppesen subscriber, you can order a "trip-kit" for an unusual trip, with no renewal service. They are very fast at fulfilling this type of order. I use Jeppesen myself. I tried the NOS bound books, but I don't like them as much as the Jepp service.

Jeppesen charts come in their own loose-leaf binders, and are updated weekly. (Jeppesen-Sanderson photo)

Some plotters warp and crack. This one won't. From Sporty's. (Sporty's photo)

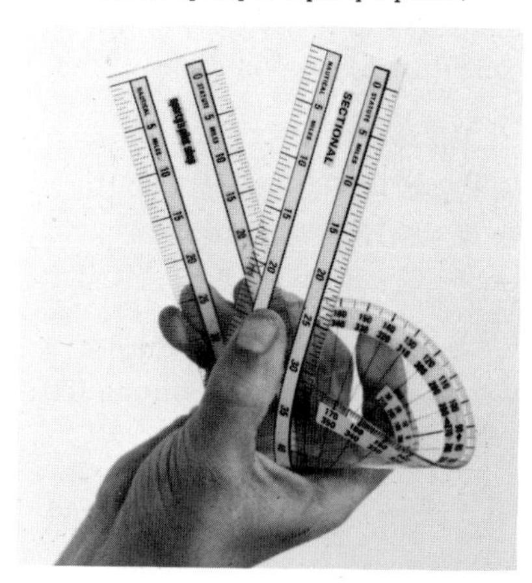

AirChart Co. publishes a bound chart service that reproduces the key NOS airport charts and offers them once a year, with a monthly "write-it-in-yourself" update service. These may be ordered through **AirChart Co., 13376 Beach Avenue, Venice, CA 90291, telephone (213) 822-1996.** This service is quite suitable for people who only do a little IFR flying, and may be used in conjunction with the NOS enroute charts.

The Canada Air Pilot is the Canadian equivalent of the NOS approach charts and enroute chart service. This may be ordered through AOPA (see above) or **Aeronautical Information Services, Canadian Air Transportation Administration, Ministry ot Transport, Ottawa K1A, 0N8, Canada.**

Airport directories

The best airport directory is "Airports USA," available from **AOPA, Box 5800, Washington, DC 20014, telephone (301) 654-0500.** It is issued once a year, free to members and $10 otherwise. It lists airports by state and gives all kinds of useful information you don't get elsewhere, such as the names and manufacturer/gasoline affiliations of the airport dealers, nearby hotels, and where the nearest Flight Service Station is located. Runway diagrams, taken from the NOS approach charts, are included in a section at the front.

Jeppesen **(Box 3279, Englewood, CO 80112, telephone (303) 779-5757)** publishes the "J-Aid," which gives airport diagrams and data, organized on a state-by-state basis. It is a loose-leaf directory, updated twice a year, and also includes most of the information contained in the "Airman's Information Manual" (AIM, see below), plus pertinent FARs and more.

Airguide Publications, Inc. (1207 Pine Avenue, Long Beach, CA 90813, telephone (213) 437-3210) publishes the "Flight Guide," a loose-leaf, small format, VFR airport directory, which is revised at regular intervals.

The "Airman's Information Manual" (AIM) includes a 7-volume airport/facility directory, which is issued every 56 days by **NOS, Riverdale, MD 20840, telephone (301) 436-6990.** Alaskan and Hawaiian supplements are also available. The *Basic Flight Information and ATC Procedures* section is issued twice a year. The *International Flight Information Manual* is issued annually; it helps general aviation flyers plan trips outside the US.

State aeronautical agency charts and directories

Remember how you used to be able to get free road maps at your local gas station? The aviation equivalent is the state aeronautical chart, published by many state agencies. See page 237 for the various state agencies' addresses. Here is a breakdown of the states that issue charts and/or directories. Most of these are free, but if they are not, the price is shown:

State	Chart	Directory
Alabama	Yes	No
Arizona	Yes	No
Arkansas	Yes	No
California	Yes ($2.50)	No
Connecticut	No	Yes
Georgia	Yes	No
Hawaii	No	Yes
Illinois	Yes	Yes
Indiana	Yes ($2.25)	No
Iowa	Yes ($3.00)	Yes
Kansas	Yes	Yes
Kentucky	Yes	Yes
Maryland	Yes	Yes
Massachusetts	No	Yes
Michigan	Yes ($1.00)	Yes ($3.00)
Minnesota	Yes	No
Mississippi	Yes	No
Missouri	Yes	Yes
Montana	Yes ($2.50)	Yes
Nebraska	Yes	Yes
New Hampshire	No	Yes
New Jersey	No	Yes
New Mexico	Yes	No
North Carolina	Yes	No
North Dakota	Yes	Yes
Ohio	No	Yes

Jepesen's J-Aid, a useful way to keep ahead of the game. (Jeppesen-Sanderson photo)

Oregon	Yes	Yes
Pennsylvania	Yes ($1.30)	Yes ($1.30)
Puerto Rico	Yes	Yes
South Carolina	Yes	No
Texas	No	Yes
Utah	Yes	Yes
Virginia	Yes	Yes
Wisconsin	Yes	Yes ($2.50)
Wyoming	Yes	Yes

Other flight guides

AOPA's Flight Department publishes some excellent guides that concentrate on specific needs. They may be obtained from **AOPA, Box 5800, Washington, DC 20014, telephone (301) 654-0500:**

USA

- Alaska
- Chicago
- Florida
- New York City
- Washington DC

Foreign

- Bahamas
- Bermuda
- Canada
- Central America
- Latin America
- Mexico
- Transatlantic and Europe
- West Indies

AOPA also publishes a pilot's guide to customs procedures.

The **Bahamas Ministry of Tourism (Box 523850, Miami, FL 33152, telephone (305) 442-4860)** offers a free air-navigation chart to the Bahamas.

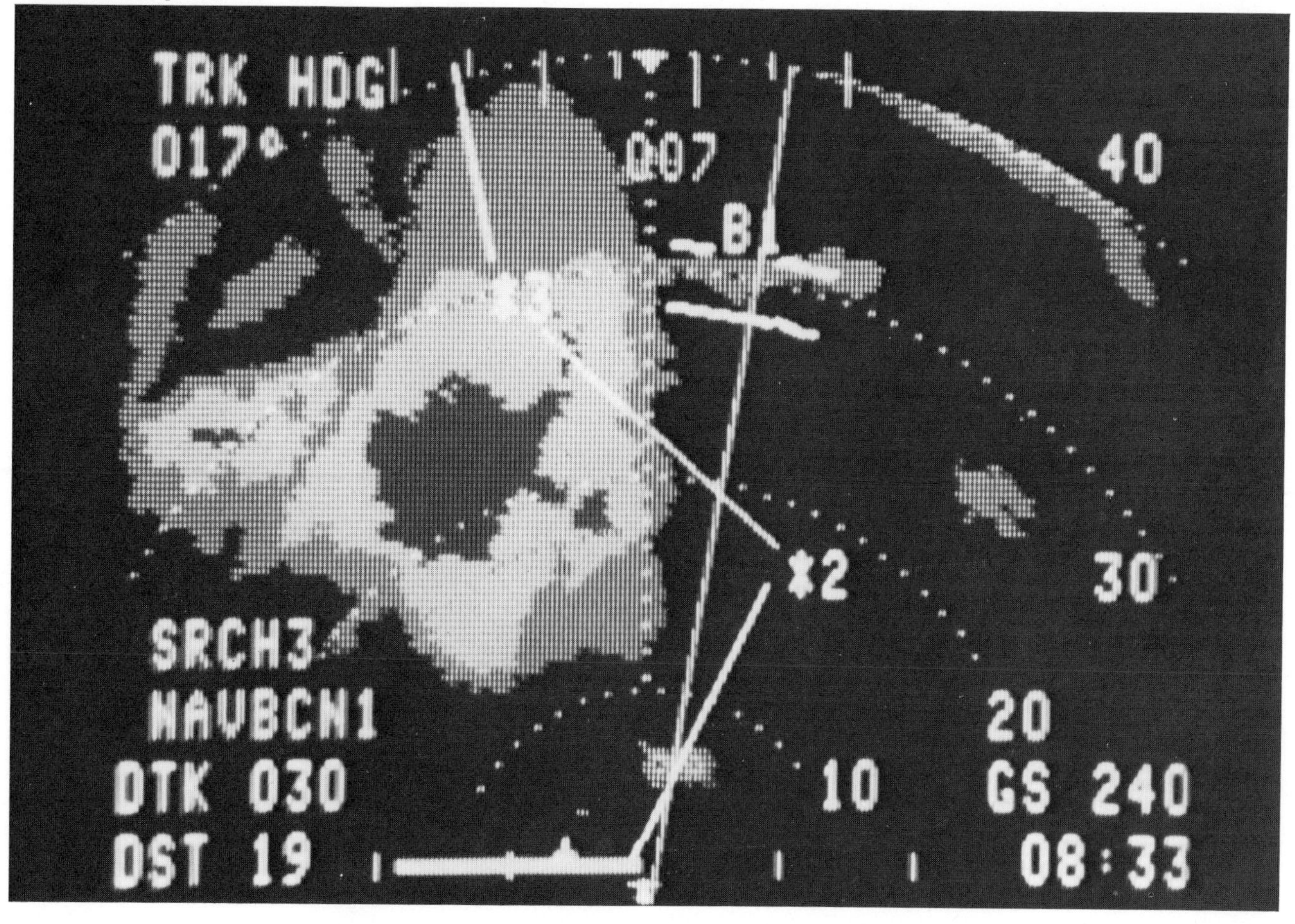

Who needs charts? Bendix color radar allows you to superimpose your RNAV routing right on the radar scope, and the same screen will show all your routing details. (Bendix photo)

The Mexican government publishes a free booklet "Fly to Mexico," which can be obtained from any Mexican Tourist Office or Consulate, or from **Direccion General De Aeronautica Civil, Departmento De Transporte Aero Internacional, Mexico City, Mexico, telephone (905) 519-6996, Telex 177-1097.** The best guide to flying in Mexico—*Mexico Flight Manual*—is published free by the **Texas Aeronautics Commission, Box 12607, Capitol Station, Austin, TX 78711, telephone (512) 475-4768.**

The Canadian government publishes a free booklet *Air Tourist Information Canada,* which gives some sparse information about visiting our neighbor to the north—I say sparse because it leaves out such important data as how to get IFR charts, where the customs airports are, and such. This booklet can be picked up at many U.S. Flight Service Stations near the border, or ordered from **Canadian Government Office of Tourism, 240 Sparks Street, Ottawa, Ontario K1A 0H6, Canada, telephone (613) 996-4610.** A good source of information about flying in Canada is the **AOPA Flight Department** (see above) or **Canadian Owners & Pilots Association, Box 734, Ottawa K1P 5S4, Canada, telephone (613) 236-4901.** Canadian charts, published in 1:500,000 and 1:1,000,000 scales, may be obtained from **Canada Map Office, 615 Booth Street, Ottawa, Ontario K1A 0E9, Canada, telephone (613) 998-3865,** as well as from AOPA and COPA.

Weather

Aviation weather services are provided by the **National Oceanic and Atmospheric Administration (NOAA) of the Department of Commerce, 6010 Executive Boulevard, Rockville, MD 20852, telephone (301) 655-4000.** The **National Weather Service (NWS)** is located at **8060 13th Street, Silver Spring, MD 20910, telephone (301) 427-7675.**

Aviation weather *forecasts* are prepared by numerous Weather Service Forecast Offices, and consist of area, terminal, route, and winds-aloft forecasts and weather synopses, produced several times a day. Hundreds of stations also report *actual* weather observations every hour and on special occasions, such as when the weather is changing rapidly. These observations are made available through U.S. Weather Bureaus (USWB), FAA Flight Service Stations (FSS), Pilots Automatic Telephone Weather Answering Service (PATWAS), En Route Flight Advisory Service (EFAS), Transcribed Weather Broadcasts (TWEB), and Scheduled Weather Broadcasts (SWB).

To find the local telephone number for aviation weather information, look in the white pages of your phone book under "U.S. Government—Transportation Department—Federal Aviation Administration—Flight Service Station or PATWAS."

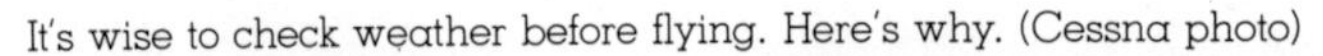

It's wise to check weather before flying. Here's why. (Cessna photo)

Daily TV show

Every weekday, a TV show called *AM Weather* is transmitted over many Public Broadcast System stations. This is broadcast generally at 6:45 AM and 7:45 AM, but the local time in your area may vary. Here is a list of stations carrying the show at press time:

■ ALABAMA

City	Station	Channel
Birmingham	WBIQ	10
Demopolis	WIIQ	41
Dozier	WDIQ	2
Florence	WFIQ	36
Huntsville	WHIQ	25
Louisville	WGIQ	43
Mobile	WEIQ	42
Montgomery	WAIQ	26
Mt Cheana	WCIQ	7

■ ARIZONA

City	Station	Channel
Phoenix	KAET	8

■ ARKANSAS

City	Station	Channel
Arkadelphia	KETG	9
Fayetteville	KAFT	13
Jonesboro	KTEJ	19
Little Rock	KETS	2

■ DELAWARE

City	Station	Channel
Wilmington	WHYY	12

"AM Weather" man Mike Tomlinson at work. (Maryland Center for Public Broadcasting photo)

■ CALIFORNIA

City	Station	Channel
Fresno	KMTF	18
Huntington Beach	KOCE	50
Los Angeles	KCET	28
San Diego	KPBS	15
San Jose	KTEH	54
San Mateo	KCSM	60

■ COLORADO

City	Station	Channel
Denver	KRMA	6
Pueblo	KTSC	8

■ FLORIDA

City	Station	Channel
Gainesville	WUFT	5
Jacksonville	WJCT	7
Miami	WLRN	17
Orlando	WMFE	16
Pensacola	WSRE	23
Tampa	WEDU	3
Tampa	WUSF	16

■ GEORGIA

City	Station	Channel
Athens	WGTV	8
Atlanta	WETV	30
Chatsworth	WCLP	18
Cochran	WDCO	15
Columbus	WJSP	28
Dawson	WACS	25
Pelham	WABW	14
Savannah	WVAN	9
Waycross	WXGA	8
Wrens	WCES	20

■ IDAHO

City	Station	Channel
Boise	KAID	4
Moscow	KUID	12
Pocatello	KBGL	10

■ KANSAS

City	Station	Channel
Topeka	KTWU	11
Wichita	WPTS	8

■ KENTUCKY

City	Station	Channel
Ashland	WKAS	25
Bowling Green	WKGB	53
Covington	WCVN	54
Elizabethtown	WKZT	23
Hazard	WKHA	35
Lexington	WKLE	46
Louisville	WKMJ	68
Louisville	WKPC	15

City	Station	Channel
Madisonville	WKMA	35
Morehead	WKMR	38
Murray	WKMU	21
Owenton	WKON	52
Pikeville	WKIP	22
Somerset	WKSO	29

■ **LOUISIANA**

City	Station	Channel
Baton Rouge	WLPB	27
Monroe	KLTM	13
New Orleans	WYES	12
Shreveport	KLTS	24

■ **MAINE**

City	Station	Channel
Augusta/Lewiston	WCBB	10

■ **MASSACHUSETTS**

City	Station	Channel
Boston	WGBH	2
Springfield	WGBY	57

■ **MARYLAND**

City	Station	Channel
Annapolis	WAPB	22
Baltimore	WMPB	67
Hagerstown	WWPB	31
Salisbury	WCPB	28

■ **MICHIGAN**

City	Station	Channel
Alpena	WCML	6
Detroit	WTVS	56
East Lansing	WKAR	23

City	Station	Channel
Grand Rapids	WGVC	35
Marquette	WMNU	13
Mt. Pleasant	WCMU	14
University Center	WUCM	19

■ **MINNESOTA**

City	Station	Channel
Appleton	KWCM	10
Austin	KAVT	15
Duluth	WDSE	8
St. Paul	KTCA	2

■ **MISSISSIPPI**

City	Station	Channel
Acherman	WMAB	2
Booneville	WMAE	12
Bude	WMAU	17
Inverness	WMAO	23
Jackson	WMAA	29
McHenry	WMAH	19
Oxford	WMAV	18
Rose Hill	WMAW	14

■ **MISSOURI**

City	Station	Channel
Kansas City	KCPT	19
St. Louis	KETC	9
Springfield	KOZK	21

■ **NEBRASKA**

City	Station	Channel
Alliance	KTNE	13
Bassett	KMNE	7
Hastings	KHNE	29
Lexington	KLNE	3

City	Station	Channel
Lincoln	KUON	12
Merriman	KRNE	12
North Platte	KPNE	9
Norfolk	KXNE	19
Omaha	KYNE	26

■ **NEW HAMPSHIRE**

City	Station	Channel
Berlin	WEDB	40
Durham	WENH	11
Hanover	WHED	15
Keene	WEKW	52
Littleton	WLED	49

■ **NEW JERSEY**

City	Station	Channel
Camden	WNJS	23
Montclair	WNJM	50
New Brunswick	WNJB	58
Trenton	WNJT	52

■ **NEW MEXICO**

City	Station	Channel
Albuquerque	KNME	5
Portales	KENW	3

■ **NEW YORK**

City	Station	Channel
Binghamton	WSKG	46
Buffalo	WNED	17
Garden City	WLIW	21
New York City	WNYC	31
Plattsburgh	WCFE	57
Rochester	WXXI	21
Schenectady	WMHT	17

City	Station	Channel
Syracuse	WCNY	24
Watertown	WNPE	16
Watertown	WNPI	18

■ **NORTH CAROLINA**

City	Station	Channel
Asheville	WUNF	33
Chapel Hill	WUNC	4
Columbia	WUND	2
Concord	WUNG	58
Greenville	WUNK	25
Linville	WUNE	17
Wilmington	WUNJ	39
Winston-Salem	WUNL	26

■ **NORTH DAKOTA**

City	Station	Channel
Fargo	KFME	13
Grand Forks	KGFE	2

■ **OHIO**

City	Station	Channel
Bowling Green	WBGU	57
Cincinnati	WCET	48
Cleveland	WVIZ	25
Columbus	WOSU	34
Dayton	WPTD	16
Oxford	WPTO	14
Portsmouth	WPBO	42
Toledo	WGTE	30

■ **OKLAHOMA**

City	Station	Channel
Cheyenne	KWET	12
Eufala	KOET	3

City	Station	Channel
Oklahoma City	KETA	13
Tulsa	KOED	11

■ **PENNSYLVANIA**

City	Station	Channel
Bethlehem	WLVT	39
Erie	WQLN	54
Hershey	WIFT	33
Philadelphia	WHYY	12
Scranton/Wilkes Barre	WVIA	44
University Park	WPSX	3

■ **SOUTH CAROLINA**

City	Station	Channel
Allendale	WEBA	14
Beaufort	WJWJ	16
Charleston	WITV	7
Columbia	WRLK	35
Florence	WJPM	33

City	Station	Channel
Greenville	WNTV	29
Rock Hill	WNSC	30
Sumter	WRJA	27

■ **SOUTH DAKOTA**

City	Station	Channel
Aberdeen	KDSD	16
Brookings	KESD	8
Eagle Butte	KPSD	13
Lowry	KQSD	11
Martin	KZSD	8
Pierre	KTSD	10
Rapid City	KBHE	9
Vermillion	KUSD	2

■ **TENNESSEE**

City	Station	Channel
Knoxville	WSJK	2
Lexington	WLJT	11
Memphis	WKNO	10

On the gauges, on oxygen, in ice, in the soup. That's weather flying. (Cessna photo)

■ **TEXAS**

City	Station	Channel
Corpus Christi	KEDT	16
Dallas	KERA	13
Houston	KUHT	8
Killeen	KNCT	46
Lubbock	KTXT	5

■ **UTAH**

City	Station	Channel
Provo	KBYU	11

■ **VERMONT**

City	Station	Channel
Burlington	WETK	33
Rutland	WVER	28
St. Johnsbury	WVTB	20
Windsor	WVTA	41

■ **VIRGIN ISLANDS**

City	Station	Channel
St Thomas	WTJX	12

■ **VIRGINIA**

City	Station	Channel
Harrisonburg	WVPT	51
Richmond	WCVE	23
Roanoke	WBRA	15
Roanoke/Norton	WSVN	47

■ **WASHINGTON**

City	Station	Channel
Pullman	KWSU	10

■ **WEST VIRGINIA**

City	Station	Channel
Grandview/Beckley	WSWP	9
Huntington	WMUL	33

■ **WISCONSIN**

City	Station	Channel
Green Bay	WPNE	38
La Crosse	WHLA	31
Madison	WHA	21
Menomonie/Eau Claire	WHWC	28
Park Falls	WLEF	36
Wausau	WHRM	20

For more information about the "AM Weather" show, contact the producers at **Maryland Public Broadcasting, Owings Mills, MD 21117, telephone (301) 356-5600.** A booklet is available for $1 from the same address that explains the weather symbols used on the show.

Continuous weather radio program

The U.S. Weather Service broadcasts continuous weather information (not particularly aviation-oriented) on a VHF frequency (162.55 mHz) in cities throughout the country. A special radio is needed to receive it. There are

"AM Weather" man Dale Bryan points out precipitation. (Maryland Center for Public Broadcasting photo)

various radios available. One, the Regency Weather Alarm Monitoradio (about $45), can receive a warning signal as well. This signal is transmitted by the Weather Service to alert listeners to special weather warnings being transmitted. There are several other weather-broadcast radios, and these may be obtained from **Sporty's Pilot Shop, Clermont County Airport, Batavia, OH 45103, telephone (513) 732-2411; Aircraft Components, Inc., 700 North Shore Drive, Benton Harbor, MI 49022, telephone (800) 253-7261, (616) 925-8861 in MI,** or from your local **Radio Shack** store.

Weather monitoring equipment

Both Sporty's and Aircraft Components, Inc., mentioned above, offer a variety of weather monitoring devices such as rain gauges, barometers, anenometers and so on. Both offer the Taylor Windscope Weather Station, which shows the wind speed and direction, barometric pressure, outside air temperature and the maximum and minimum

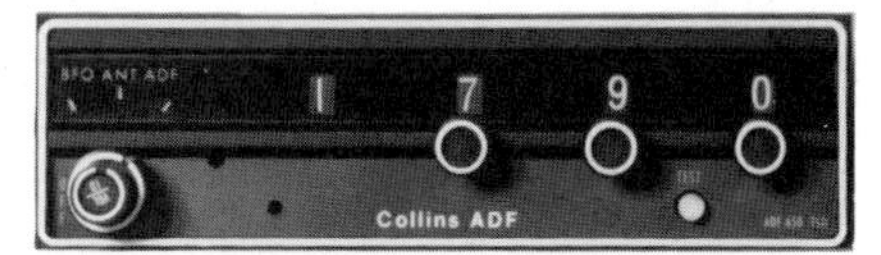

You can listen to aviation "TWEBs" (transcribed weather broadcasts) on your ADF, but not the VHF continuous weather broadcasts. (Collins photo)

temperatures. This wall-mounted display is useful for use in small airport pilot-lounges and costs about $470.

Books and materials

A free pamphlet, *A Pilot's Guide to Aviation Weather Services*, is available from Flight Service Stations; it shows the various services available and explains how to read the weather sequences as they come off the teletype. Two FAA Advisory Circulars are #00-6A *Aviation Weather* ($4.55), and 00-45A *Aviation Weather Services* ($3.00). These may be obtained from the **Superintendent of Documents, US Government Printing Office, Washington, DC 20402,** or from **Government bookshops.**

Other books are:

- *Fair Weather Flying* by Richard L. Taylor (Macmillan)
- *Flying the Weather Map* by Richard L. Collins (Delacorte)
- *Pilot's Handbook of Weather* by Guerney and Skiera (Aero)
- *Pilot's Weather* by Ann Welch (Fell)
- *Pilot's Weather Guide* by Lindy Boyes (Modern)
- *Weather Flying* by Robert N. Buck (Macmillan)

There is a device called a *Pilot's Weather Computer*, created by Samuel Kramer. It is offered by **Telex Communications, Inc. (9600 Aldrich Avenue South, Minneapolis, MN 55420, telephone (612) 884-4051)** through their dealers.

Another device is called *Pocket Weather Trends*. It's a sort of sliderule that lets you compare the pictures of cloud formations and so on depicted on it with what you see outside, and thus helps you draw some conclusions about the weather for the next 12 to 36 hours. It is available from **Sporty's** (see address above) at $2.95.

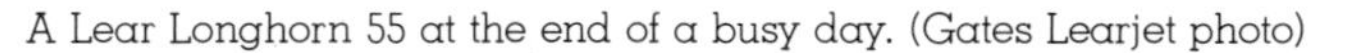

A Lear Longhorn 55 at the end of a busy day. (Gates Learjet photo)

DOUGLAS
DC 3
FORD
TRI-MOTOR
NORTHROP
ALPHA

P A R T F I V E

Aerobuff Stuff

AIR SHOWS

AIRSHIPS

BOOKS ABOUT FLYING

CONCORDE

MAGAZINES

MOVIES ABOUT FLYING

AVIATION MUSEUMS

RESTAURANTS MADE OUT OF AIRPLANES

TOURS

The National Air and Space Museum in Washington, DC. (Tim Foster photo)

Air Shows

There are two types of air show to be found in America. One is the trade show, designed to increase the sales of the industry's wares; the other is the straight air show, designed to show the public some fancy flying.

The trade shows come every year around the same time. The first show of the season is the Reading Show, officially called the "National Maintenance and Operations Meeting." This runs for four days, Tuesday through Friday, at Reading Municipal Airport in Reading, Pennsylvania, in early June. It used to be an annual event, but every odd year it clashed with the Paris Air Show, which the same exhibitors attending Reading wanted to attend, making it very difficult for them to handle both shows. So finally Reading bowed, and now the show is held only in even years—1980, 1982, etc. The big hangar and a nearby building are turned into huge exhibit areas, and the ramp space is covered with aircraft on display. You may be able to save a dollar on the admission charge by registering in advance through the Aircraft Owners & Pilots Association: **AOPA, Washington, DC, 20014, telephone (301) 654-0500.** If you're an AOPA member, they'll tell you about it. The **Reading Show** people publish a visitors' guide, which may be obtained from them at **Box 1201, Reading, PA 19603, telephone (215) 375-8551.**

I really like the Reading Show,

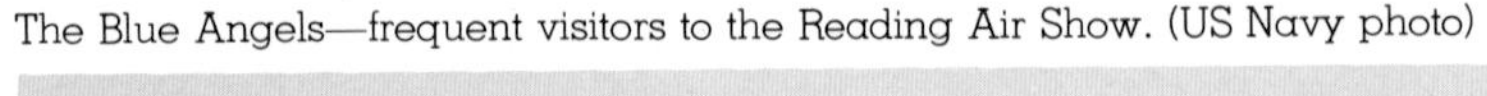

The Blue Angels—frequent visitors to the Reading Air Show. (US Navy photo)

because it has good exhibits of aeronautical paraphernalia—it's like a live version of *The Aviator's Catalog*. Flying in can be a little hectic. If you have to go IFR, forget Reading. Fly in to Allentown, PA, and take the shuttle bus—which takes about 45 minutes. If you're going in VFR, get there as early as possible—before 8 AM if you can. If you arrive near 11 AM or after, you'll probably have to hold for half an hour or more. The field closes each day during the afternoon for flying demonstrations.

The next show, and the best for true buffs, is the Experimental Aircraft Association Fly In at Oshkosh, WI. This is held in early August each year in permanent facilities at Whittman Field in Oshkosh. This is the biggest fly-in event in the world, with literally 10,000 airplanes attending with their human partners. Attendance by people comes to over 300,000 during the show, which runs more than a week. Here you will see hundreds of homebuilt aircraft, warbirds, antiques, and fascinating new designs. There is a large exhibit area, which is on a less sophisticated level than that at Reading—no color radar or executive jets, but lots of real grassroots stuff like do-it-yourself kits, parts, propellers, tires—you name it. They even have an aviation flea market, where you can pick up genuine aeronautical junk. The finest in cottage industry is also represented in the shape of such artifacts as lamps made from seashells, stones with faces painted on them, ceramics of the most garish design, every conceivable T-shirt message, and custom-

The Thunderbirds at work. (USAF photo)

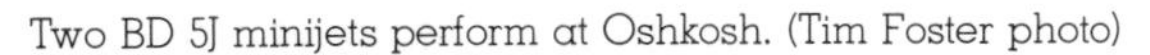

Two BD 5J minijets perform at Oshkosh. (Tim Foster photo)

made plastic nametags. The major light aircraft manufacturers have finally recognized the importance of the Oshkosh show, and they are all represented with their latest wares.

Flying in is an experience you won't want to miss. Oshkosh becomes the world's busiest airport during the show, and the ingenuity of the controllers in handling the constant flow of arriving and departing traffic, plus aircraft doing fly-bys, is astounding. If you're going in IFR, I suggest you fly in to Green Bay and rent a car. Getting an IFR departure out of Oshkosh can be an experience to try the patience of the most stoic individual.

Details about The Oshkosh show can be obtained from the **Experimental Aircraft Association, Box 229, Hales Corners, WI 53130, telephone (414) 425-4860.**

The next trade show is the National Business Aircraft Association convention, which moves around the country each year, and is held in September. In 1979 it was held in Atlanta, and Kansas City in 1980, Anaheim in 1981, St Louis in 1982, Dallas in 1983, and Atlanta again in 1984.

This show is the top-of-the-line in trade shows, since it caters principally to the business-flying market, which means jets, helicopters, inertial navigation systems, and big-money operations. The show is open to the public (admission is about $10), and is principally an exhibit presentation, with meetings and workshops. Aircraft can be seen at the nearby airport, but there is no flying display. This is where to go to see the most exotic equipment available, but don't get the idea that it's all jets. Even the top-of-the-line singles and twins are covered. **National Business Aircraft Association, One Farragut Square South, Washington, DC 20006, telephone (202) 783-9000.**

The Aircraft Owners and Pilots Association holds its annual Plantation Party and Trade Exhibit each year around October, moving around the country to a resort area each time. Las Vegas and Miami are two frequent locations. This is more of a fraternal event than a trade show. There are banquets and "outings for the ladies," as well as golf and tennis tournaments. There is an exhibit area, which ranks behind those at Reading, Oshkosh, or the NBAA in scope, but which is still interesting. Airplanes are to be seen at the airport or in the hotel parking lot in some places, but there is no flying display. When you attend the major events, you get a free bolo tie with an AOPA Plantation Party badge on it. Each year the badge is changed, with the new date and location inscribed. It is mindboggling to see people walking around with more than a dozen of these enamel badges strung out on their ties! The AOPA Plantation Party is a popular social event. **AOPA, Washington, DC 20014, telephone (301) 654-0500.**

For helicopter people, the Helicopter Association of America holds an annual convention in January of each year. This moves around various western resort locations each year (Tucson, AZ, Las Vegas, NV, Anaheim, CA) and features exhibits and demonstrations of choppers. **The Helicopter Association of America, Inc., 1156 15th Street NW, Washington, DC 20005, telephone (202) 466-2420.**

There are two international shows that should be mentioned. One is the Paris Air Show, held in alternate (odd) years at Le Bourget Airport, Paris, France in June, and the other is the Farnborough Air Show, held at the Royal Aircraft Establishment, Farnborough, England, in September in the other alternate (even) years. These shows cater much more to the overall aviation industry than American shows do. You'll see the latest military aircraft from all over the world being demonstrated, and the exhibit areas tend to be technical. I grew up at the Farnborough Air Show, since I went to school on the downwind leg of the main runway. Thus I was exposed to all of Britain's new aircraft innovations at an early age. The Farnborough Show used to be confined to British developments, but with the internationalization of the aircraft industry in Europe, it was opened to a wider spectrum of participants a few years ago. For information about the Paris Air Show, contact **French Trade Shows, 1350 Avenue of the Americas, New York, NY 10019, telephone (212) 582-4960.** For information about the Farnborough Show, contact **British Information Services, 845 Third Avenue, New York, NY 10022, telephone (212) 593- 2258.**

There are two major shows in Canada each year. The Abbotsford International Show is held at Abbotsford, British Columbia (near Vancouver) in August. Contact **Abbotsford Air Show Society, Box 361, Abbotsford, British Columbia, V2S 4N9, Can-**

ada, telephone (604) 859-9211. There is a flying display-only show held at the end of the annual Canadian National Exhibition in Toronto, with activities on the days immediately before and including Labor Day. This is strictly a gee-wiz air show, designed to impress the folks attending the biggest annual "state fair"—the Canadian National Exhibition.

There are hundreds of air shows held throughout the United States each year at local airports, often with the support of the big military aerobatic exhibit teams, The Blue Angels (U.S. Navy) or The Thunderbirds (U.S. Air Force). These are listed in the calendar section of most aviation magazines and are publicized locally. These local shows have more of the old barnstorming—wing walking, parachute jumping, Ford Trimotor rides, aerobatics, and the inevitable "crazy student" flight, where the student ostensibly gets the Piper Cub into the air when the instructor isn't with him and then demonstrates how badly an airplane can be flown and still be landed and walked away from.

If you are thinking of putting on your own air show, I suggest you get a copy of the *EAA Air Show and Fly-In Manual,* published at $2.85 by the **Experimental Aircraft Association, Box 229, Hales**

The Blue Angels. (US Navy photo)

The Christen Eagle puts on a good show at Oshkosh. (Christen photo)

The *Columbia*. (Goodyear photo)

Corners, WI 53130, telephone (414) 425-4860. The FAA also has an Advisory Circular, #91-45A *Airshow Waivers*. This may be obtained free from **U.S. Department of Transportation, Publications Section M443.1, Washington, DC, 20590.**

Airships

Ever since the Hindenberg tragedy in 1937, airships have been *aeronautica-non-grata* for travelers. After their discontinuation as passenger carriers, they continued to be used for years by the U.S. Navy, but they were phased out in 1962. This leaves us with Goodyear and its blimps, which is a good place to be left.

Goodyear uses four airships as its "aerial ambassadors," with three of these being based in the United States and one in Europe. They are named the *America*, the *Columbia*, the *Mayflower*, and the *Europa*. The *Mayflower* was severely damaged in a storm in 1979 and has since been rebuilt. Goodyear has built over 300 lighter-than-air craft since 1917, which makes it the most experienced airship manufacturer in the world. Its present fleet is used for public service, public relations, advertising, and research.

Goodyear employs five pilots, a ground crew of sixteen people, and one public relations person per ship. Each blimp is supported by a crew bus, which also serves as a communications and command post, and a large truck for maintenance services, as well as night sign and TV equipment facilities and a small van.

The Goodyear blimps are based in Pompano Beach, Florida, Los Angeles, Houston, and Italy. From these bases they rove around the United States and Europe, acting as camera ships for major events, such as the Super Bowl football game or the Tall Ships celebration in New York Harbor in 1976, and generally participating in community activities.

Going for a ride

If you're where the blimp is, it might be possible to get a ride. I went up in one once in Toronto, during Canada's Centennial celebrations in 1967. I liked it very much, and I am surprised by the lack of acceptance of blimps as a means of transportation. Why aren't there hundreds of them? In a blimp you feel perfectly safe. The takeoff is a little like an airplane, and it climbs at a fairly steep angle, with its twin engines at full power. The blimp shape supplies some lift. When we were airborne, the pilot throttled down both engines and we just sat there, in almost complete silence, about 1,000 feet above the ground. The feeling was one of serenity and security. The blimp, in fact, doesn't even have seat belts. The most distinctive aspect of the controls is the huge wheel, about the size of a hula hoop, mounted vertically beside the pilot. This is, in effect, a gigantic trim wheel, and is used to handle the elevators. The other controls are conventional. The instrumentation is basic IFR airplane stuff, plus a few extras referring to the helium pressure and so on. I asked the pilot about how they operate the blimp. He said that they didn't operate in high winds or icing conditions, but the blimps were perfectly capable of IFR flight—in fact the blimp pilot's license includes IFR privileges.

Goodyear offers public rides in its Florida-based blimp between November and April. Reservations are required and can be made on a day-to-day basis by calling **(305) 946-8300** in Pompano Beach. No advance reservations are accepted. A ride costs $7.50 for adults and $5 for children under 12. Each blimp carries only six passengers, so early action is called for if you want to go up.

No, it's not a wheelchair. It's the pilot's seat in the Goodyear blimp. (Goodyear photo)

Other airships

Various moves are underway to make use of airships in commercial air transportation. Plans have been offered for a giant airship to carry oil instead of transporting it by pipeline. In Britain, **Aerospace Developments (19-21 Newbury Street, London EC1A 7HU, England, telephone 01-606-5981)** has developed and flown the AD-500 non-rigid airship, using new construction techniques. Orders for this have come in from South America, and the ship is also being evaluated by the U.S. Navy. The airship is covered in great detail in the February 24, 1979, issue of **Flight International, Dorset House, Stamford Street, London SE1 9LU, England, telephone 01-261-8070.** It is priced at about $1.5 million. Airships soon won't be just big Goodyear advertisements, in my opinion.

Associations

The American Airship Association calls itself an "information center for all aerostatic vehicles," and may be reached at **5300 Westboard Avenue, Washington, DC 20016, telephone (301) 656-9592. The Lighter-Than-Air-Society, Inc.,** is at **1800 Triplett Boulevard, Akron, OH 44306, telephone (216) 794-1321.**

Books

The best book on the subject of airships is *The Blimp Book*, by George Larson, with photos by George Hall and Baron Wolman, published by Squarebooks. This contains some beautiful photography and everything you ever wanted to know about the Goodyear blimp.

Books About Flying

Aviation has been well-served in literature. There is a book on just about every subject. However, you can go into many bookstores and find that they don't even have a section dedicated to aviation books. So they're to be found under "Sports" (you'll find football, fishing, and flying all bunched together), or "Military" (where you'll find *Stick and Rudder* alongside *German Tanks of the Desert Campaign* and *Torture Techniques of the Gestapo*), or sometimes, not much more intelligently, under "Automotive" (*Chilton's Guide to the Volkswagen* and *Used Car Price Guide* alongside *The Aircraft Owner's Handbook*). Just once in a while, you'll find a real aviation section, with all the aviation books grouped together. Unless it's a big store, you'll probably be disappointed by the selection. So that brings us to two other sources—the aviation book specialists and the aviation book clubs.

The specialist stores offer catalogs, and some have retail outlets. The most complete aviation book listing and store is that of **Sky Books International, 48 East 50th Street, New York, NY 10022, telephone (212) 688-5086.** Other catalogs are available from **Aero Publishers, Inc., 329 West Aviation Road, Fallbrook, CA 92028, telephone (714) 728-8456; Aviation Book Company, 1460 Victory Boulevard, Glendale, CA 91202, telephone (213) 240-1771; Pan American Navigation Service, 16934 Saticoy Street, Box 9046, Van Nuys, CA 91409, telephone (213) 345-2744.**

If you get to England, a specialist store well worth a visit is **Beaumont's Aviation Bookshop, 656 Holloway Road, London N19 3PD, England, telephone 01-272-3630.** They also offer a catalog(ue), with mail-order service to anywhere in the world.

Aviation book clubs

There are three aviation book clubs:

Flying Book Club
Box 454
Hightstown, NJ 08520

Jeppesen Aviation Book Club
Box 2007
Latham, NY 12111

Modern Aviation Library
Blue Ridge Summit, PA 17214

Books

Here is a list of aviation books, organized by subject matter. Information given is title—author—publisher, in that order.

Basic flying

- *Airman's Information Manual* by Aero Publishers
- *America's Flying Book* by Flying Magazine (Scribners)
- *Anyone Can Fly* by Jules Bergman (Doubleday)
- *Back to Basics* by Flying Magazine (Van Nostrand Reinhold)
- *Basic Guide to Flying* by Fillingham (Hawthorne)
- *Cleared for Takeoff* by Gordon Stokes (Scribners)
- *Cockpit Navigation Guide* by Don Downie (Modern Aircraft)
- *Computer Guide* by Frank Kingston Smith (Modern Aircraft)
- *FAA Flight Test Guide—Private Pilot, Airplane* by Aero Publishers
- *Fair Weather Flying* by Richard L. Taylor (Macmillan)
- *Federal Aviation Regulations* by Aero Publishers
- *Federal Aviation Regulations and Airman's Information Manual* by Aero Products Research
- *Federal Aviation Regulations for Pilots* by Pan American Navigation Service
- *Flight Briefing for Pilots* (4 volumes) by N. H Birch and A. E. Bramson— (Ziff Davis)
- *Flight Maneuvers Manual* by Haldon (Aviation Book Company)
- *Flight Manual* by Jeppesen
- *Flight Review* by Jeppesen
- *Flying off the Pavement* by Grindle (Aviation Book Company)
- *Flying Safely* by Richard Collins (Delacorte Press)
- *Flying Wisdom* by Flying Magazine (Van Nostrand Reinhold)
- *How to Fly Floats* by Jay Frey (Edo Corporation)
- *I Learned About Flying From That!* by Flying Magazine (Delacorte/Eleanor Friede)
- *Invitation to Flying* by Robert S. Hunter (Ziff-Davis)
- *The Joy of Learning to Fly* by Gay Dalby Maher (Delacorte/Eleanor Friede)
- *Manual of Flight* by Aero Products Research
- *The New Private Pilot* by Pan American Navigation Service
- *A Pilot's Digest of FAA Regulations* by John Nelson (Modern Aircraft)
- *Pilot Error* by Flying Magazine (Van Nostrand Reinhold)
- *Pilot's Guide to Flight Emer-*

gency Procedures by N. H. Birch and A. E. Bramson (Doubleday)

- *Pilot's Handbook of Navigation* by J. C. Elliott and G. Guerny (Aero Publishers)
- *Pilot's Handbook of Weather* by G. Guerney and J. Skiera (Aero Publishers)
- *Pilot Instruction Manual* by FAA (Doubleday)
- *The Pilot's Night Flying Handbook* by Len Buckwalter (Doubleday)
- *A Pilot's Survival Manual* by Paul Nesbitt, Alonzo Pond, and William Allen (Van Nostrand Reinhold)
- *Pilot's Weather* by Ann Welch (Frederick Fell)
- *Pilot's Weather Guide* by Lindy Boyes (Modern Aircraft)
- *Private Pilot Complete Programmed Course* by Aero Products Research
- *Private Pilot Course (Mach 1)* by Jeppesen
- *Private Pilot's Guide* by Larry Reithmaier (Aero Publishers)
- *Private Pilot Manual* by Jeppesen
- *Private Pilot Study Guide* by Leroy Simonson (Aviation Book Company)
- *Private Pilot Study Pak* by Jeppesen
- *Safety After Solo* by John Hoyt (Pan American Navigation Service)
- *Stick and Rudder* by Wolfgang Langewiesche (McGraw-Hill)
- *The Student Pilot's Flight Manual* by W. K. Kershner (Iowa State University Press)
- *Weather Flying* by Robert Buck (Macmillan)
- *Weekend Pilot* by Frank Kingston Smith (Random House)
- *Your Pilot's License* by Joe Christy and Clay Johnson (Modern Aircraft)

Advanced flying

- *Advanced Pilot's Flight Manual* by W. K. Kershner (Iowa State University Press)
- *Agricultural Aviation Guide* by Alan Hoffsommer (Modern Aircraft)
- *Airline Pilot Course (Mach 4)* by Jeppesen
- *Airline Transport Pilot* by Zweng (Pan American Navigation Service)
- *As the Pro Flies* by John Hoyt (McGraw-Hill)
- *ATP-Airline Transport Pilot* by K. T. Boyd (Iowa State University Press)
- *Commercial Course (Mach 3)* by Jeppesen
- *Corporate Flying* by Jack L. King (Aviation Book Co.)

- *FAA Flight Test Guide* by Commercial Pilot, Airplane (Aero Publishers)
- *FAA Flight Test Guide—Flight Instructor* by Aero Publishers

Jeppesen's Private Pilot kit. (Jeppesen-Sanderson photo)

- *FAA Flight Test Guide—Instrument Pilot, Airplane* by Aero Publishers
- *FAA Flight Test Guide—Multiengine, Airplane* by Aero Publishers
- *Flight Instructor's Manual* by W. K. Kershner (Iowa State University Press)
- *Flying IFR* by Richard Collins (Delacorte Press)
- *Flying the Weather Map* by Richard Collins (Delacorte Press)
- *Fly the Wing* by Jim Webb (Iowa State University Press)
- *Guide to Air Traffic Control* by Robert T. Smith (Modern Aircraft)
- *Handling the Big Jets* by Davies (Pan American Navigation Service)
- *IFR Pocket Simulator Procedures* by Henry Culver (Flight Information Publications)
- *Instrument Flight Manual* by W. K. Kershner (Iowa State University Press)
- *Instrument Flying* by Richard L. Taylor (Macmillan)
- *Instrument Flying Guide* by Robert T. Smith (Modern Aircraft)
- *Instrument Pilot's Guide* by Larry Reithmaier (Aero Publishers)
- *Instrument Rating Course (Mach 2)* by Jeppesen
- *Instrument Rating Manual* by Jeppesen
- *Instrument Rating Study Pak* by Jeppesen
- *The Instrument Rating* by Zweng (Pan American Navigation Service)
- *Mountain Flying* by Sparky Imeson (Airguide Publications)
- *Multiengine Airplane Rating* by T. M. Smith (Pan American Navigation Service)
- *Nav/Com Guide for Pilots* by Don Downie (Modern Aircraft)
- *Your Jet Pilot Rating* by F. Bunyan (Modern Aircraft)

Sport flying

- *Aerobatics* by Neil Williams (Doubleday)
- *America's Soaring Book* by Flying Magazine (Scribner)

- *The Art and Technique of Soaring* by Richard Wolters (McGraw Hill)
- *The Complete Beginners Guide to Soaring and Hang Gliding* by Norman Richards (Doubleday)
- *The Complete Book of Sky Sports* by Linn Emrich (Collier)
- *The Complete Soaring Pilot's Handbook* by Ann and Lorne Welch & Frank Irving (McKay)
- *The Encyclopedia of Hot Air Balloons* by Paul Garrison (Drake Publishers)
- *Hang Flight: Flight Instruction Manual for Beginner and Intermediate Pilots* by Joe Adleson and Bill Williams (Eco-Nautics)
- *Hang Gliding* by Ross R. Olney (Putnam)
- *Hang Gliding* by Dorothy Schmitz (Crestwood)
- *Hang Gliding: Basic Handbook of Skysurfing* by Dan Poynter (Parachuting Publications)
- *Hang Glider's Bible* by Michael A. Markowski (Tab)
- *The Hang Gliding Book* by William Bixby (McKay)
- *Hang Gliding Handbook: Fly Like a Bird* by George Siposs (Tab)
- *Hang Gliding: The Flyingest Flying* by Don Dedera (Northland)
- *Hang Gliding: Rapture of the Heights* by Lorraine M. Doyle (Aviation Books)
- *Hang Gliding: Riding the Wind* by Otto Penzler (Troll)
- *Hang Gliding and Soaring: a Complete Introduction to the Newest Way to Fly* by James E. Mrazek (St Martin)
- *Modern Aerobatics and Precision Flying* by Harold Krier & Bill Sweet (Modern Aircraft)
- *Parachuting: The Skydivers Handbook* by Dan Poynter (Parachuting Publications)
- *Parachuting Manual and Log* by Dan Poynter (Parachuting Publications)
- *Primary Aerobatic Flight Training* by A. C. Medore (Aviation Book Company)
- *Roll Around a Point* by Duane

Cole (Aviation Book Company)

- *Sailplanes and Soaring* by James E. Mrazek (Hawthorn)
- *Skysurfing* by Eddie Paul (Modern Aircraft)
- *Soaring Guide* by Peter M. Bowers (Modern Aircraft)

Non-piloting activities

- *Aero Mechanics Questionnaire* by Ralph Rice (Aero Publishers)
- *The Aircraft Mechanic's Shop Manual* by Larry Reithmaier (Palomar)
- *Airframe Mechanic's Manual* by Zweng (Pan American Navigation Service)
- *Flight Engineers Manual* by Zweng (Pan American Navigation Service)
- *How to be a Flight Stewardess or Steward* by Johni Smith (Pan American Navigation Service)
- *Powerplant Mechanics' Manual* by Zweng (Pan American Navigation Service)

Airplane data

- *Aircraft Dope and Fabric* by Ruth Spencer (Modern Aircraft)
- *The Aircraft Owner's Handbook: Everything You Need to Know About Buying, Operating and Selling an Aircraft* by Timothy R. V. Foster (Van Nostrand Reinhold)
- *Buying and Owning Your Own Airplane* by James Ellis (Iowa State University Press)
- *Cessna Guidebook* by Mayborn & Pickett (Flying Enterprises)
- *Custom Light Planes, A Design Guide* by William A. Welch (Modern Aircraft)

- *Design for Flying* by David Thurston (McGraw-Hill)
- *Every Pilot's Guide to Aviation Electronics* by John Ferrara (El-Jac Publishing Co.)
- *Flying the Old Planes* by Bill Tallman (Doubleday)
- *The Great Planes* by James Gilbert (Grosset & Dunlap)
- *Guide to Homebuilts* by Peter M. Bowers (Modern Aircraft)

- *The Illustrated Encyclopedia of the World's Commercial Aircraft* by William Greene and Gordon Swanborough (Crescent)
- *Janes All the World's Aircraft* by John W. R. Taylor (Franklin Watts)
- *Janes Pocket Books, Commercial Transport Aircraft* by John W. R. Taylor (Collier)
- *Janes Pocket Books, Homebuilt Aircraft* by Taylor (Collier)
- *Janes Pocket Books, Light Aircraft* by Taylor (Collier)
- *Janes Pocket Books, Major Combat Aircraft* by Taylor (Collier)
- *Janes Pocket Books, Military Transport and Training Aircraft* by Taylor (Collier)
- *Janes Pocket Books, Record Breaking Aircraft* by Taylor (Collier)
- *Janes Pocket Books, Research and Experimental Aircraft* by Taylor (Collier)
- *Jet Aircraft Engines* by Irwin E. Treager (Modern Aircraft)
- *The Piper Cub Story* by James Triggs (Modern Aircraft)
- *Lightplane Construction and Repair* by Snyder and Welch (Modern Aircraft)
- *Modern Lightplane Engines* by John L. Nelson (Modern Aircraft)
- *The Observer's Basic Civil Aircraft Directory* by Green/Swanborough (Frederick Warne)
- *The Observer's Book of Aircraft* by William Green (Scribners)
- *The Observer's Military Aircraft Directory* by Green/Swanborough (Frederick Warne)

- *The Observer's World Airlines Directory* by Green/Swanborough (Frederick Warne)
- *A Practical Guide to Airplane Performance and Design* by Donald R. Crawford (Crawford)
- *Restoration of Classic/Antique Planes* by Don Dwiggins (Modern Aircraft)
- *Ryan Guidebook* by Mayborn (Flying Enterprises)
- *The Single-Engine Beechcrafts* by Joe Christy (Modern Aircraft)
- *The Single-Engine Cessnas* by Joe Christy (Modern Aircraft)
- *The Single Engine Pipers* by John Duncan (Modern Aircraft)
- *Stearman Guidebook* by Mayborn and Bowers (Flying Enterprises)
- *Those Incomparable Bonanzas* by Larry Ball (McCormick-Armstrong Company)

Miscellaneous

- *AOPA Airports USA* by Aircraft Owners & Pilots Association (AOPA)
- *AOPA Handbook for Pilots* by AOPA
- *AOPA Flight Guides—AOPA: Alaska, Bahamas, Bermuda, Canada, Central America, Florida, Latin America, Mexico, New York City, Transatlantic & Europe, Washington, DC, West Indies*
- *Aviation Insurance Explained* by J. C. White (Aviation Book Company)
- *Bax Seat* by Gordon Baxter (Ziff-Davis)
- *The Best of Flying* by Flying Magazine (Van Nostrand Reinhold)
- *Fate is the Hunter* by Ernest K. Gann (Simon & Schuster)
- *The Flier's World* by James Gilbert (Random House)
- *Flight Through The Ages* by C. H. Gibbs-Smith (Thomas Y.Crowell Company)
- *The Lore of Flight* by Tré Tryckare (Time-Life Books)
- *Pilot and Aircraft Owner's Legal Guide* by White (Aviation Book Company)
- *Preventive Maintenance for Pilots and Aircraft Owners* by Crane (AMFI)
- *The Private Pilot's Dictionary and Handbook* by Kirk Polking (Arco Publishing)
- *Private Pilot's Blue Book* by Dwyer (Stein & Day)
- *Skywriting* by James Gilbert (St. Martin's Press)
- *The Spirit of St. Louis* by Charles Lindbergh (Ballantine)

There are hundreds more titles. I have excluded the enthusiasts' books for example—all those tomes that review in great detail the camouflage of each variety of JU 88 that served in World War II and so on. As I said earlier, for the most complete book catalog, write to **Sky Books International, Inc., 48 East 50th Street, New York, NY 10022.**

FAA advisory circulars

See also page 235. The FAA publishes a number of Advisory Circulars for a price (and many more at no charge). The principal charged-for books, organized by subject matter, are listed below. They may be bought at Government Book Stores, FBOs, and through the various aviation book outlets already mentioned.

FLYING TECHNIQUES

Title	Circular Number
■ *Student Pilot Guide*	AC 61-12G
■ *Private Pilot's Handbook of Aeronautical Knowledge*	AC 61-23A

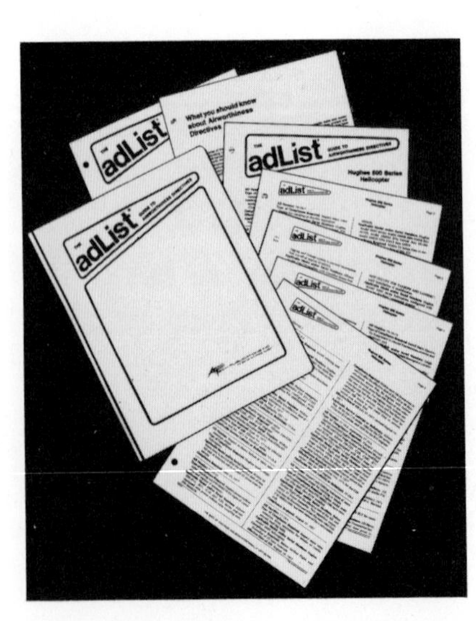

"adList" is a service for people seeking information about FAA airworthiness directives for particular aircraft. It is available on a plane-by-plane basis from AeroTech Publications, Box 528, Old Bridge, NJ 08857, telephone 201-679-5151.

"adLog" is an aircraft maintenance record-keeping system, customized for your aircraft. It may be ordered from AeroTech (see address above).

- *Private Pilot (Airplane) Flight Training Guide* AC 61-2A
- *Terrain Flying* AC 91-15
- *Pilot Transition Courses, Complex Single-Engine and Light Twins* AC 61-9B
- *Private and Commercial Pilot Refresher Courses* AC 61-10A
- *Basic Helicopter Handbook* AC 61-13A
- *Instrument Flying Handbook* AC 61-27B
- *Flight Training Handbook* AC 61-21
- *Flight Instructor's Handbook* AC 61-16A

WRITTEN TEST GUIDES

- *FAR Written Test Guide for Private, Commercial and Military Pilots* AC 61-34B
- *Private Pilot, Airplane* AC 61-32A
- *Instrument Pilot* AC 61-8C
- *Commercial Pilot, Airplane* AC 61-71
- *Private and Commercial Pilot (Rotorcraft-Helicopter)* AC 61-73
- *Flight Instructor, Airplane* AC 61-72
- *Flight Instructor, Instrument* AC 61-70
- *Flight Instructor, Rotorcraft-Helicopter* AC 61-74
- *Flight Instructor, Glider* AC 61-75
- *Ground Instructor, Basic-Advanced* AC 143-1D
- *Ground Instructor, Instrument* AC 143-2B
- *Airline Transport Pilot, Airplane* AC 61-18D
- *Airline Transport Pilot, Helicopter* AC 61-42A
- *Flight Engineer Written* AC 63-1B
- *Flight Navigator Written* AC 63-2A
- *Aircraft Dispatcher* AC 65-4B
- *Parachute Rigger Certification Guide* AC 65-5A
- *Private and Commercial, Gyroplane* AC 61-60
- *Private and Commercial, Glider* AC 61-61

MECHANICS

- *Airframe and Powerplant Mechanics Certification Guide* AC 65-2C
- *Airframe and Powerplant Mechanics Handbook* AC 65-9
- *Airframe and Powerplant Mechanics Certification Guide* AC 65-11A
- *Airframe and Powerplant Mechanics Powerplant Handbook* AC 65-12
- *Airframe and Powerplant Mechanics Airframe Handbook* AC 65-15

FLIGHT TEST GUIDES

- *Private Pilot, Airplane* AC 61-54A
- *Instrument Pilot, Airplane* AC 61-56A
- *Commercial Pilot, Airplane* AC 61-55A
- *Flight Instructor, Airplane* AC 61-58
- *Airline Transport Pilot, Airplane* AC 61-77
- *Type Rating, Airplane* AC 61-57A
- *Private and Commercial, Helicopter* AC 61-59
- *Instrument Pilot, Helicopter* AC 61-64

MISCELLANEOUS

- *Aviation Weather* AC 00-6A
- *Aviation Weather Services* AC 00-45
- *Forming and Operating a Flying Club* AC 00-25
- *Personal Aircraft Inspection Handbook* AC 20-9
- *Pilot's Weight and Balance Handbook* AC 91-23
- *Basic Glider Criteria Handbook* AC 21-3
- *Medical Handbook for Pilots* AC 67-2
- *Guide to Drug Hazards in Aviation Medicine* AC 91-11-1

FAA safety materials

The FAA publishes various safety pamphlets and booklets. *Exam-O-Grams* covering both VFR and IFR subjects are designed to correct misconceptions that pilots display in written exams, flight tests, and accident and violation reports. They are free, and may be ordered from **Flight Standards National Field Office, AFS-590, Box 25082, Oklahoma City, OK 73125.** The FAA will put you on their mailing list for future *Exam-O-Grams* if you ask them.

Other FAA safety pamphlets are produced for handing out free at aviation safety seminars, and are also available at FAA General Aviation District Offices (GADOs) and your local FSS. Here are some of the titles:

- *Always Leave Yourself an Out*
- *Safety Guide for Private Aircraft Owners*
- *Tips on Engine Operations in Small General Aviation Aircraft*
- *Tips on How to Use the Flight Planner*
- *Tips on Winter Flying*

FAA catalog

The Guide to FAA Publications contains a complete listing of FAA booklets, along with ordering information. This may be obtained from **FAA Public Inquiry Center, APA-430, 800 Independence Avenue SW, Washington, DC 20591, telephone (202) 426-8058.**

Concorde

How could I refuse the opportunity to fly across the Atlantic in the Concorde? In the spring of 1979, I had to go over to England for a few days on business, and as a study in contrasts, as well as to save a few bucks to help pay for the Concorde return, I took the Laker Skytrain over from New York to London-Gatwick. This was a relatively painless six-hour trip in a DC 10, and it cost a paltry $139. Sir Freddie Laker has set up a good thing for the masses. The check-in procedures are a little different from those expected by experienced airline travelers, but everything is clear and straightforward.

In anticipation of writing this account, I arranged with British Airways well in advance to obtain permission to ride up front in the jump seat on the Concorde. This was set up through New York with no difficulty. Perhaps the fact that I was paying the $998 one-way fare myself helped!

Concorde pax = VVIPS

At London's Heathrow Airport everything is very much organized to accomodate the Concorde passenger, who is treated as a *Very* Important Person. Even the access signs on the way in specifically mention Concorde when directing you to Terminal 3. Concorde passengers have their own check-in counter with minimal "queueing."

As you proceed through the massive international terminal, you follow the special Concorde signs and again feel important as you go through passport control. Here the Concorde passengers go through a special gate that serves to '"queue-jump" them to the head of the line. Then to the Concorde lounge, where stewards serve complimentary champagne and hors-d'oeuvres. On my arrival in the lounge, I found my advance planning had paid off, because I was immediately introduced to Captain Brian Calvert, Concorde check captain, who was waiting for me. It turned out that our pilot, Captain Peter Duffey, was to undergo a standard route check on this flight. This was very beneficial to me, since on the flight he called out every move so the check pilot would know what he was doing and why, and thus I was kept informed of all activities as they happened.

Concorde is normally crewed by three—two pilots and a flight engineer. There are two jump seats available, one behind the captain, today occupied by Captain Calvert, and one behind everyone, in the "tunnel" to the main cabin. This was to be my seat for most of the flight. I was given a headset and an oxygen/emergency-exit briefing by Captain Calvert. Two of the 38,050-pound-thrust Bristol Olympus engines were started before push back, and the other two were started after the tractor had disconnected.

Departure procedures

Prior to take off, the nose was tilted down five degrees and the windshield visor was retracted (for landings, the nose is tilted down 12-1/2 degrees). The aircraft is equipped with triple inertial navigation systems (INS), which are programmed on the ground with large plastic cards featuring an array of symbols op-

The Concorde is on its way. (British Airways photo)

tically read by the computer, thus entering all the waypoint information for the entire flight. The zero fuel weight on push back was 86.6 metric tons (190,520 pounds), the fuel load was 94.7 metric tons (208,340 pounds), and the ramp weight was 181.34 metric tons (398,940 pounds or just under 200 tons).

Captain Duffey carried out a formal and detailed pre-takeoff briefing for the crew, spelling out exactly who was to do what in the event of various emergencies. Our initial clearance was for departure off runway 28R, with a 250-knot/4,000-foot restriction until 7 DME from the Woodley VORTAC. The takeoff numbers were V1—195 knots, VR—197 knots, V2—219 knots, and a first-segment climb of 250 knots. Although 28R is 3,900 meters long (12,800 feet), the end of the runway was quite close as we lifted off and climbed away. The surface wind was 210 degrees at 20 knots, and there was a fair amount of low-level turbulence below the 5,000-foot strato-cu scattered cloud layer. I was surprised by how much this affected the Concorde—it seemed to

Concorde Captain Peter Duffy in the left seat and Captain Brian Calvert in the right. (British Airways photo)

bounce around just as much as my Comanche would! Shortly after takeoff, the Captain called out "three, two, one—noise!" and the flight engineer pulled the throttles back a long way to produce the noise abatement segment of the climb.

Very soon we received our clearance to continue climbing to FL 280. As we went through 4,500 feet, First Officer Christopher Orlebar selected "normal nose," and the visor was extended over the windshield. This resulted in an immediate and significant cabin noise reduction. "Quiet, isn't it?" remarked Captain Duffey. "So much nicer," replied the F/O. "We still have the passengers strapped in?" asked the Captain. "I was just waiting until we were out of this chop." "I think we can undo them now." "Right, sir."

Going supersonic

We accelerated to a climb speed of Mach .95 and held this all the way to FL 280. As we approached that altitude, we received a further clearance to cruise climb between FL 490 and FL 570. Now came the time for the acceleration to supersonic flight. Concorde is the only airline equipped with afterburners, and these are used on takeoff and for the "accel." You can really detect when the afterburners are applied. You feel a definite acceleration, but no vibration or buffeting. We went through Mach 1 at FL 290 and accelerated gradually as we climbed, reaching Mach 2 as we went through FL 520, where our CAS was 530 knots. Our rate of climb was under 2,000 fpm for this segment, so it took about fifteen minutes to get up there. From there on Mach 2 was maintained, and the altitude was allowed to fluctuate at this constant speed. In fact we cruised at somewhere around FL 590.

Small canapé trays were passed up to the crew—caviar, paté, and shrimp. All partook voraciously—these were obviously part of their staple diet. I was invited to go back to the cabin and enjoy the lunch, which I did. I found myself sitting next to another pilot, a man from California, and the owner of a turbo-Twin Comanche. Needless to say, we found much to talk about, and I later arranged for him to come up to the flight deck and observe proceedings. Lunch consisted of Dom Perignon, lamb cutlet, and fresh fruit salad. This was followed by Remy Martin Napoleon Cognac. I spent about 45 minutes in the cabin, and then we heard Captain Duffey announce we were about 180 miles south of Newfoundland—we'd only been up a little over two hours! So I returned to the front office and sat in the regular jump seat.

The big slowdown

Very soon, we commenced deceleration, which consisted of a substantial power reduction and a noticeable lunge forward of the body as our personal inertia reacted to the slow-down. Our objective was to hit Mach 1 about 43 miles east of Hyannis, Mass. Captain Calvert told me the sonic-boom "footprint" was about 35 miles ahead of the aircraft, so they go subsonic a few miles more than that from land. He added that after the Concordes had been in service a while someone had noticed that the Mach number digits to the right of the decimal point indicated miles to go to Mach 1. In other words, if you're at Mach 1.85 and decelerating, you have 85 miles to go before being subsonic. This is handy, because, believe it or not, Concorde doesn't have VNAV. As we slowed down, we also descended—at about 4,000 fpm. Soon we were down to a more normal altitude—FL 390, and we were talking to Boston Center. The controller pointed us out to a Delta flight so that their passengers could see us, and the Delta pilot asked, "How fast are you going, Concorde?" "We're just down to Mach .95." "It looks really beautiful up there," replied Delta.

Registration idiosyncrasy

I asked what the registration of this Concorde was. "G-N81AC," replied Captain Calvert. "No, I don't mean the serial number, I mean the British registration," I replied. "That's it. When this aircraft operates for Braniff between Washington and Dallas, we simply get up on the tail and peel off the G and the aircraft becomes N81AC—American registered—so that it can legally fly for an American airline!" I was amazed. The British government had changed its registration procedures to accomodate this requirement, since normal British registrations start with G and are followed by four letters—never numbers (except some flight-test aircraft). "The Braniff flights are operated by Braniff crews, aren't they?" "Yes." "But they don't operate supersonically, do they?" "No." "Sort of makes them eunuchs, doesn't it?" "Yes." Ah, well.

The approach

Soon we were working Kennedy Approach. Our call sign "Speedbird Concorde 171 Heavy" was a mouthful, and was often abbreviated to Speedbird 171. The nose was tilted down for the approach, and again Captain Duffey gave a formal crew briefing, listing the tasks of each member in the event of an emergency. We were to carry out an ILS on 31R. The autopilot was flying the airplane and soon we were firmly on the ILS, descending at about 1,000 fpm. The INS depicted the wind, and the HSI (horizontal situation indicator) showed a small bug indicating the wind direction, a useful feature. Touchdown occurred at our exact scheduled arrival of 1015 AM EDT—an hour before we left! Our actual off-to-on time was 3:27. Concorde uses a lot of runway, but it has immensely powerful brakes and these slowed us down effectively. As we taxied in, we saw British Airways' London-bound Concorde lifting off runway 31 L and making a sharp turn to the left. Soon we were at the gate, and I made profuse gestures of gratitude to the crew, who had been most friendly and helpful all the way. I showed them a picture of my Comanche and suggested that they should keep out of my way, since I'm always flying in the New York TCA. "And you keep out of *our* way!" "I shall, indeed. Thank you, gentlemen, for a fabulous flight. You have now ruined me for future flights, because this is the *only* way to go!" "Yes, now they've found out we won't eat their children, it has got rather popular!" Flying at one mile every 2.7 seconds—1,175 knots or 1,350 mph, is certainly an experience I shall never forget. I hope to repeat it soon.

Magazines

There are dozens of aviation magazines in North America. The largest of these is *Flying*, with a circulation of about 400,000. *Flying* started out as *Popular Aviation* in August, 1927, and changed its name in 1940. It is published by Ziff-Davis, monthly, and is edited by Richard Collins, son of Leighton Collins, who founded and edited *Air Facts* in the 1930s. (*Air Facts*, alas, is no more). **Flying Magazine, One Park Avenue, New York, NY 10016.**

Next largest in circulation is a magazine you can only get if you belong to the Aircraft Owners and Pilots Association—*AOPA Pilot*. Since AOPA has about 230,000 members, *Pilot*'s circulation is the same. It, too, is a monthly, and is edited by Ed Tripp. **AOPA Pilot, Box 5800, Washington, DC 20014.**

A magazine that has changed in character somewhat over the years is *Air Progress*. Twenty years ago, it was mostly an aviation enthusiast's magazine, but it is now more of a flying enthusiast's book. However, it still carries news of military aviation, something you seldom read about in the other general-aviation publications. Its circulation is about 150,000, and it is edited by Keith Connes. *Air Progress* moved its editorial offices from New York to Los Angeles a few years ago when it was sold, making it the largest circulation aviation magazine coming from the west coast. Like most aviation magazines, *Air Progress* is a monthly. **Air Progress, 7950 Deering Avenue, Canoga Park, CA 91304.**

Two other monthly, west-coast magazines appeared on the newsstands at about the same time a few years ago—*Plane and Pilot* and *Private Pilot*. *P&P* has a circulation of about 90,000 and is edited by Bill Cox. **Plane and Pilot, 606 Wilshire Boulevard, Santa Monica, CA 90401.** *Private Pilot* sells about 90,000 copies and is edited by Dennis Shattuck. **Private Pilot, Box 4030, San Clemente, CA 92672.**

The publishers of *Private Pilot* also publish *Aero*, a monthly that is sent free to most aircraft owners. This business of free sub-

"I say, chaps, anyone for a spot of tea?" Compare this photo of a BOAC crew in a Boeing Stratocruiser with that of the Concorde flight deck on page 190. Note morse key for wireless operator, wooden chair for flight engineer, and steely, confident gaze/profile of ace Captain. The crew is wearing their "tropicals." (British Airways photo)

scriptions is called "controlled circulation." It means, in theory, that only the *right people* get the magazine—those to whom it is aimed. And *Aero* is aimed at registered aircraft owners. Its editorial content has improved considerably since it changed hands. It has a circulation of about 70,000, and is, like its sister publication, edited by Dennis Shattuck. **Aero, Box 4030, San Clemente, CA 92672.**

There are three monthly magazines that cater to the business and professional pilot—*Business and Commercial Aviation*, *Professional Pilot*, and *Flight Crew*. *B/CA* is published by Ziff-Davis (the people who publish *Flying)* and is the best of the three, in my opinion. It is edited by John Alcott and is distributed by controlled circulation to owners of late-model, high-performance, single-engine aircraft, as well as to those operating twins, helicopters, and turbine equipment. It has a circulation of about 50,000. **Business and Commercial Aviation, Hangar C 1, Westchester County Airport, White Plains, NY 10604.**

Professional Pilot has controlled distribution to pro pilots numbering 32,000. It is edited by Murray Queen Smith. It is a unique magazine, offering interesting insights into business, air taxi, and commuter aircraft operations. **Professional Pilot, West Building, Washington National Airport, Washington, DC 20001.**

A new magazine *Flight Crew* is also controlled to the higher-performance aircraft operators and specializes more in articles about pilot technique and operational problems than news reports of the industry. **Flight Crew, 137 East 36 Street, New York, NY 10016.**

There are several publications that specialize in carrying aircraft advertising—to the complete exclusion of editorial material, in some cases. The leader, and everybody's bible, is *Trade-A-Plane*, which has a tabloid newspaper format and is issued three times a month. The only editorial material is a not-very-funny cartoon on the front page—usually commenting on how some intrepid birdman could not survive without his *Trade-A-Plane*. It is printed on yellow paper and will be seen in virtually every FBO's back office and hangar in the country. **Trade-A-Plane, Crossville, TN 38555.**

Ziff-Davis gets into the act again with *AC Flyer*, a glossy monthly magazine consisting solely of ads—almost all in full color. It is controlled to aircraft owners, but after Z-D bought it a year or so ago, they have been going after subscriptions. **AC Flyer, 444 Brickell Avenue, Miami, FL 33131.**

There is an assortment of other "airplane ad magazines," some of which are:

North Atlantic Aviation
Box 186
Brookfield, CT 06804

Air List Ads
Box 1196
Fairfield, NJ 07006

AMALSA Buyer's Digest
201 Commerce Drive
Fort Collins, CO 80521

Aviation Buyers' Guide
Box 538
Paoli, PA 19301

US Aircraft Dealers
2929 Convair Road
Memphis, TN 38116.

These, again, are usually controlled to aircraft owners.

There are a few aviation newspapers that have fairly regional circulations. *Flight Line Times* comes out weekly and is published by the same people who do *North Atlantic Aviation*. **Flight Line Times, Box 186, Brookfield, CT 06804; General Aviation News, Box 1094, Snyder, TX 79549,** comes out every other Monday, and **Aviation News, 13920 Mount McClellan Avenue, Reno, NV 89506,** is a monthly.

The main newsmagazine of the aviation industry is *Aviation Week*, published by McGraw-Hill. It comes out as often as its name implies and is edited by William Gregory. It suffers from a fascination with tanks and weapon systems that have nothing to do with airplanes, and has very poor coverage of the general-aviation market. However, it is good on things like production progress on the Boeing 767, what airline just bought three used DC 9s from United, and military activity. **Aviation Week, 1221 Avenue of the Americas, New York, NY 10020.**

An important source of news to the business flying community is another Ziff-Davis epic—*Business Aviation*, which is a weekly newsletter published in Washington, DC. It carries very limited advertising and few illustrations—just news—and it costs $145 a year. **Business Aviation, 1156 15th Street NW, Washington, DC 20005.**

Aviation Consumer, edited by Richard Weeghman, is another unique magazine that comes out twice a month—one which I recommend to anyone who owns an airplane or is thinking hard about buying one. It carries no advertising and is a very consumer-oriented publication that goes into many facets of flying that you never hear about in the glossy monthlies. You will learn more about airplanes and their little problems from *Aviation Consumer* than from any other single source. One especially useful feature is a thorough used-airplane report, where the history of the aircraft in question is discussed, including all of the variations and modifications that have occurred, laced heavily with owner comments. **Aviation Consumer, 1111 East Putnam Avenue, Riverside, CT 06878.**

Aviation Convention News is a large-format magazine that is published at the times of the various major air shows, such as the Reading Show, the National Business Aircraft Association Show, the Helicopter Association Show, and so on. It is controlled to high-performance general-aviation aircraft operators and is published daily at the NBAA show as a handout. **Aviation Convention News, Pan AM Building, Teterboro, NJ 07608.**

The ubiquitous Ziff-Davis also publishes "show dailies" at most major shows. The only way you can get these is to attend the show.

Flying publishes *Flying Annual* early each year, which is a useful guide to currently available hardware, presented with several interesting articles about the current state of the industry and where everything is going. **Flying Annual, One Park Avenue, New York, NY 10016.**

Plane and Pilot puts out a number of newsstand-only magazines with titles like *Used Planes Guide*, *Flight Test Annual*, and *Learn to Fly*. These are basically rehashes of old *Plane and Pilot* articles.

There are several specialty magazines. *Sport Aviation* is published by the Experimental Aircraft Association, and you get it monthly with membership. **Sport Aviation, Box 229, Hales Corners, WI 53130.**

Rotor and Wing International, Box 1790, Peoria, IL 61656, deals mostly with helicopter operations.

Air Line Pilot, 1625 Massachusetts Avenue, Washington, DC 20036, goes to members of the Air Line Pilots Association and deals mostly with key airline-pilot problems, such as loss-of-license insurance, seniority, and pensions.

There are more, but the above should keep you going for a while!

Movies about Flying

Some of the finest—and some of the worst—movies ever made were about flying. One of my pet bugs is to watch a movie that shows people getting into a Boeing 727, a Boeing 707 taking off, a DC 8 landing, and then the people getting out of a DC 9—all in the same flight! But they're not all like that. Herewith a list of some of the best flying films:

- *Air Force* (1943. South Pacific air battles.)
- *Airport* (1970. Arthur Hailey's story, with a cast of thousands of Hollywood names. But Dean Martin as an airline pilot??? Followed by *Airport 1975* and *Airport '77*.)
- *Angels One Five* (1954. An earlier Battle of Britain epic. Also called *Hawks in the Sun*. Jack Hawkins and John Gregson.)
- *Appointment in London* (1953. Dirk Bogarde, Bryan Forbes. WW II RAF bomber squadron stuff.)
- *The Battle of Britain* (1969. Super buff's film with real Spitfires, Hurricanes, Messerschmitt 109s, Junkers 89s, and more.)
- *The Big Lift* (1950. Very well done Berlin Air Lift story, with Montgomery Clift. Paul Douglas gives a good GCA.)
- *Birds of Prey* (1973. TV movie with David Janssen. Terrific chase scenes with low-flying helicopters over the Utah desert.)
- *The Blue Max* (1966. All about the honorable German fighter pilots of World War I. George

Peppard, James Mason, and Ursula Andress. Great flying scenes, filmed in Ireland.)

- *Bombers B 52* (1957. An updated *Strategic Air Command.*)
- *Breaking the Sound Barrier* (1952. How the British tried it. Great drama. Lots of stiff upper lips.)
- *Breakout* (1975. True story about a Mexican jailbreak via helicopter. Charles Bronson. Film later inspired another jailbreak in Michigan after being shown to the inmates!)
- *The Bridges at Toko-Ri* (1954. William Holden as a reserve Navy pilot flying Panther jets in the Korean War.)
- *Capricorn One* (1978. Interesting idea about faking a flight to Mars. Terrific chase scenes between a Hughes 500 and a Sterman in the desert mountains. Bill Tallman's last film.)
- *Captains of the Clouds* (1942. Jimmy Cagney as a Canadian bush pilot who can't quite settle into the RCAF's ways.)
- *Ceiling Zero* (1935. Jimmy Cagney gets his instrument rating.)
- *The Dam Busters* (1955. How the RAF bombed some strategic and indestructible dams in the Ruhr. True story. Jolly good show.)
- *The Dawn Patrol* (1938. Stiff upper lips in WW I France. Errol Flynn and David Niven .)
- *Destination Moon* (1950. Good special effects. Dated as hell, but interesting.)
- *Devil Dogs of the Air* (1935. Rivals Jimmy Cagney and Edmond O'Brien fight it out in the clouds.)
- *Dive Bomber* (1941. What happens when you pull 6 Gs. Eroll Flynn.)
- *The Doomsday Flight* (1966. Rod Serling TV movie. Good suspense. Edmond O'Brien.)
- *Eagle Squadron* (1942. Robert Stack joins the RAF in WW II.)
- *Fighter Squadron* (1948. Edmond O'Brien. WW II action and Rock Hudson's first film.)
- *Flight Command* (1940. How to become a naval aviator, with Robert Taylor, Walter Pidgeon. Interesting planes.)
- *Flight for Freedom* (1943. Rosalind Russell as an aviatrix, loosely based on Amelia Earhart.)
- *The Flight of the Phoenix* (1966. About a group of survivors building an airplane out of their wrecked Fairchild Flying Boxcar. Paul Mantz was killed making this film.)

All dressed-up to go to see *Twelve O'Clock High!* This authentic jacket comes from Avirex's "Combat Aeronautica" collection. (Avirex photo)

- *Flying Leathernecks* (1951. John Wayne and the U.S. Marines.)
- *Flying Tigers* (1942. More John Wayne WW II heroics).
- *A Gathering of Eagles* (1963. SAC revisited. Rod Taylor, Rock Hudson.)
- *The Great Waldo Pepper* (1975. Nicely done Robert Redford epic about barnstorming. Great flying scenes.)
- *The Gypsy Moths* (1969. Burt Lancaster and skydiving.)
- *Hell's Angels* (1930. Howard Hughes' amazing air spectacle.)
- *The High and the Mighty* (1954. Ernest Gann story. John Wayne again. Good suspense about an airliner trying to make it over the Pacific with heavy engine trouble. Great theme song.)
- *Island in the Sky* (1953. Another Ernie Gann book. John Wayne and crew surviving a C 47 crash in the Arctic and how they were found.)
- *The Lost Squadron* (1932. Erich Von Stroheim. WW I air force vets take up making flying movies.)
- *The Malta Story* (1953. Alec Guiness. How they defended Malta in WW II with only three aeroplanes.)
- *Men of the Fighting Lady* (1954. Van Johnson. What it was supposed to be like on an aircraft carrier during the Korean War.)
- *Mosquito Squadron* (1970. David McCallum. WW II action at low altitudes).
- *No Highway in the Sky* (1951. A Nevil Shute story, with Jimmy Stewart as an aircraft designer and Marlene Dietrich as a movie star. Phony model airplane used, with a biplane-tail! Good story.)
- *Reach for the Sky* (1956. The story of Douglas Bader, the legless British air ace. Kenneth More.)
- *633 Squadron* (1964. Excellent flying shots of DH Mosquitos

The dastardly Terry-Thomas and his evil-minded sidekick Eric Sykes in *Those Magnificent Men in their Flying Machines.* (20th Century Fox photo)

carrying out an impossible bombing raid in a Norwegian fjord. Cliff Robertson. Much inspiration for *Star Wars.*)

- *Sky Heist* (1975. TV movie. Helicopter chase scenes.)
- *Sky Riders* (1976. Hang gliders come into their own. James Coburn.)
- *Skyjacked* (1972. Charlton Heston hijack movie.)
- *Spitfire* (1942. Leslie Howard. How the Spitfire was born. Also called *First of the Few.*)
- *The Spirit of St. Louis* (1957. Long and engrossing film about Lindbergh's Paris flight, starring Jimmy Stewart.)
- *Star Wars* (1977. Not really a flying movie, but so what? It *feels* like a flying movie!)
- *Strategic Air Command* (1955. Jimmy Stewart gives up baseball coaching to keep the peace. Lot's of B-47's and June Allyson.)
- *The Tarnished Angels* (1957. All about the 1930's air racing scene. Good flying shots of rare airplanes.)
- *Test Pilot* (1938. Clark Gable and Spencer Tracy. Very art-deco.)
- *Thirty Seconds Over Tokyo* (1944. How to fly a B-25 off a carrier. Spencer Tracy.)
- *Those Magnificent Men in Their Flying Machines* (1965. Enormous fun about a London-to-Paris air race in Edwardian days. Lots of fascinating old airplanes.)
- *Tora! Tora! Tora!* (1970. Pearl Harbor story. Realistic battle scenes.)
- *Twelve O'Clock High* (1949. Gregory Peck helps the USAAF turn the corner in latter World War II. Lots of B-17s.)
- *2001 A Space Odyssey* (1968. Super special effects.)
- *The War Lover* (1962. Steve McQueen as a hotshot B-17 pilot in WW II England.)
- *X-15* (1961. Experimental airplanes abound in this story about test pilots. Charles Bronson.)
- *Winged Victory* (1944. Edmond O'Brien. WW II Stuff.)
- *Wings* (1927. The Classic WW I air movie.)

Movies—documentary and educational

There are many sources of documentary and educational films on aviation subjects. The FAA maintains a film library, and their films may be borrowed free from the **FAA Film Service, 5000 Park Street, St. Petersburg, FL 33709, telephone (813) 541-7571.** (This is actually Modern Talking Pictures, who handle the film distribution for the FAA.) Here are some of their titles, with their length and catalog numbers (where the title is not an obvious description of the title, a brief description is given):

Aerodynamics and conditions of flight

- *Caution: Wake Turbulence*—16 mins—FA-10-70
- *Density Altitude*—29 mins—1969—FA-603A
- *How Airplanes Fly*—18 mins—1969—FA 703
- *Mountain Flying*—23 mins—FA-06-75
- *Some Thoughts on Winter Flying*—21 mins—FA-01-75

Through the miracle of film, these SB2C Helldivers live on. Which movie? One of those oldies about life on board the carriers, I guess. (US Navy photo)

- *Stable and Safe*—20 mins—1969—FA-704 (Talks to non-instrument rated pilots)
- *Stalling for Safety*—18 mins—FA-04-74
- *Takeoffs and Landings*—12 mins—FA-02-03-04-75

Aircraft

- *Flying Floats*—19 mins—FA-03-73
- *A Plane is Born*—27 mins—1968—FA-602
- *Plane Sense*—20 mins—1968—FA-807 (What to watch out for when buying a used plane)
- *Safety by the Numbers*—31 mins—1969—FA-802 (Multi-engine techniques)

Airports

- *Airports in Perspective*—15 mins—1969—FA-706
- *Airports Mean Business*—28 mins—FA-06-72
- *It Pays to Stay Open*—23 mins—1966—FA-609 (Danny Kaye talks about how round-the-clock operations at two Massachusetts airports paid off)
- *Where Airports Begin*—20 mins—FA-05-75 (How two communities successfully planned and developed their airports)

Air traffic control

- *ATC and Its Performing ARTS*—6 mins—FA-05-72 (About the automated terminal radar system)
- *First Stop to Safety*—5 mins—FA-04-72 (How the Flight Service Station will meet the needs of the future)
- *Flight 52*—15 mins—FA-02-74 (How the US National Airpsace System works)
- *The Flight Service Station*—28 mins—1969—FA-901

Aviation medicine

- *All It Takes Is Once*—25 mins—1969—FA-801 (The effects of five psychological problems on pilots)
- *Charlie*—22 mins—1967—FA-618 (Flying and drinking don't mix)
- *Disorientation*—19 mins—FA-09-73
- *Eagle Eyed Pilot*—25 mins—FA-08-71 (All about the human eye and flying)
- *Hypoxia*—16 mins—FA-08-73
- *Medical Facts for Pilots*—25 mins—FA-01-70
- *Restraint for Survival*—8 mins—1967—FA-805 (Seatbelts and shoulder harnesses)
- *Rx for Flight*—20 mins—1968 FA-606 (Basic aeromedical problems for pilots)

General aviation and flying clubs

- *Flying clubs*—20 mins—1969—FA-705
- *General Aviation—Fact or Fiction?*—15 mins—FA-01-73
- *Path to Safety*—20 mins—1967—FA-612 (Cliff Robertson as a flying instructor discussing how misjudgment can cause problems)

Navigation

- *Area Navigation*—25 mins—FA-02-70
- *Basic Radio Procedures for Pilots*—30 mins—1970—FA-902
- *Dusk to Dawn*—29 mins—FA-06-71 (Why an instrument rating is important)
- *Radar Contact*—29 mins—FA-07-72 (All about radar and transponders)

Weather

- *Air Masses and Fronts*—20 mins—1965—AP-3 (Slide/tape show—51 slides)
- *The Atmosphere*—23 mins—1965—AP-1 (Slide/tape show—58 slides)

Modern talking pictures titles:

- *General Aviation—Making the Difference*—14 mins
- *To Fly*—27 mins

NASA films

The National Aeronautics and Space Administration has an excellent selection of films. Here is a list of some of the NASA films that are available:

Films about the space program

- *The Apollo 4 Mission*—16 mins—1968—HQ-181
- *The Flight of Apollo 7*—14 mins—1968—HQ-187
- *Debrief: Apollo 8*—28 mins—1969—HQ-188
- *Apollo 9: The Space Duet of Spider and Gumdrop*—29 mins—1969—HQ-189
- *Apollo 10: Green Light for a Lunar Landing*—29 mins—1969—HQ-190
- *Eagle Has Landed: The Flight of Apollo 11*—29 mins—1969—HQ-194
- *Apollo 12: Pinpoint For Science*—28 mins—1969—HQ-197
- *Apollo 13: "Houston . . . We've Got a Problem."*—28 mins—1970—HQ-200

Apollo 15 LM and Rover with Astronaut James Irwin on the moon.

- *Apollo 14: Mission to Fra Mauro*—28 mins—1971—HQ-211
- *Apollo 15: In the Mountains of the Moon*—28 mins—1971—HQ-217
- *Apollo 16: Nothing So Hidden*—28 mins—1972—HQ-222
- *Apollo 17: On the Shoulders of Giants*—28 mins—1973—HQ-227
- *The Time of Apollo*—28 mins—1975--HQ-225
- *Apollo-Soyuz (Background)*—28 mins—1975—HQ-242
- *The Mission of Apollo-Soyuz*—28 mins—1975—HQ-256
- *Assignment: Shoot the Moon*—28 mins—1967—HQ-167
- *Blue Planet*—10.5 mins—1972—HQ-224 (An overview of the space program and its benefits)
- *The Moon—Old and New*—26 mins—1970—HQ-209

- *Research Project X-15*—27 mins—1966—HQ-79
- *Space in the 1970's—Aeronautics*—28 mins—1971—HQ-205
- *Space Shuttle*—15 mins—1976—HQ-285
- *Spaceport—USA*—23 mins—1970—HQ-210 (All about the Kennedy Space Center)

Other NASA films

- *America's Wings*—28 mins—1976—HQ-267
- *A Man's Reach Should Exceed His Grasp*—24 mins—1972—HQ-219 (The story of flight and man's reach for freedom through aviation and the exploration of space)

Here's how to order the NASA films:

If you live in:

Alaska, Arizona, California, Hawaii, Idaho, Montana, Nevada, Oregon, Utah, Washington, Wyoming

Contact NASA Education Office at:

NASA Ames Research Center
Moffett Field, CA 94035
(415) 965-5000

If you live in:

Alabama, Arkansas, Iowa, Louisiana, Mississippi, Missouri, Tennessee

Contact NASA Education Office at:

NASA George C. Marshall Space Flight Center
AL 35812
(205) 453-2121

If you live in:

Delaware, DC, Maryland, New Jersey, Pennsylvania, Vermont

Contact NASA Education Office at:

NASA Goddard Space Flight Center
Greenbelt, MD 20771
(301) 982-5042

If you live in:

Florida, Georgia, Puerto Rico, Virgin Islands

Contact NASA Education Office at:

NASA J. F. Kennedy Space Center, FL 32899
(305) 867-7113

If you live in:

Kentucky, North Carolina, South Carolina, Virginia, West Virginia

Contact NASA Education Office at:

NASA Langley Research Center
Hampton, VA 23365
(804) 827-1110

If you live in:

Illinois, Indiana, Michigan, Minnesota, Ohio, Wisconsin

Contact NASA Education Office at:

NASA Lewis Research Center
21000 Brookpark Road
Cleveland, OH 44135
(216) 433-4000

If you live in:

Colorado, Kansas, Nebraska, New Mexico, North Dakota, Oklahoma, South Dakota, Texas

Contact NASA Education Office at:

NASA Lyndon B. Johnson Space Center
Houston, TX 77058
(713) 483-3111

If you live in:

Connecticut, Maine, Massachusetts, New Hampshire, New York, Rhode Island

Contact NASA Education Office at:

National Audiovisual Center
General Services Administration
Washington, DC 20409
(301) 763-5500

Various commercial enterprises have films that may be rented or borrowed. Here is a selection:

From: **Goodyear Tire and Rubber Company**
Public Relations Film Library
1144 East Market Street
Akron, OH 44316

The Blimps—Clearly Identified Flying Objects—15 mins—1968

From: **Lockheed-Georgia Company**
Motion Picture Film Library
Zone 30, B-2 Building
Marietta, GA 30063

Wings at Work—a History of Commercial Aviation—28 mins

From: **McDonnell Douglas Corp.**
Film Distribution and Promotion
2525 Ocean Park Blvd.
Santa Monica, CA 90405

- *Ambassadors In Blue*—15 mins—1971 (All about the USAF Thunderbirds and their Phantoms)
- *The Challenge*—19 mins—1969 (A history of the USAF—WW I to Viet Nam)

The Thunderbirds on film. (USAF photo)

- *Diamond in the Sky*—14 mins—1972— (All about the USN Blue Angels and their Phantoms)
- *The Eagle at Farnborough*—9.5 mins—1974 (The F-15 Eagle visits Britain)
- *Genesis 10*—14.5 mins—1971 (How the DC-10 was developed)
- *Phantom Joins the Fleet*—22 mins—1962
- *Saga of the Skyraider*—14 mins—1967
- *Thunderbirds Premier*—19 mins—1970
- *Tradition Blue Angels*—15 mins—1969
- *The Wings of Youth*—14.5 mins—1965 (Civil Air Patrol activities)
- *To Fly!*—27 mins—1976—#31496 (The film shown at the National Air and Space Museum)

From: **Shell Film Library**
1433 Sadlier Circle West Drive
Indianapolis, IN 46239

How an Airplane Flies—56 mins—1976

Not to be undaunted, the armed services have many films available for loan at no charge. The Air Force offers films through:

Central Audiovisual Library
Aerospace Audiovisual Service
Norton AFB, CA 92409

There are at least 33 films in a series called *The Air Force Story*, which runs from *The Beginning* to *Human Factors and Space Flight*. Here are some of the other titles:

- *C-5 Galaxy—World's Largest Aircraft*—28 mins—1969—SFP 1768
- *Stepping Stones to the Stars—The Air Force Museum*—27 mins—1973—SFP 2026
- *High Flight*—3 mins—1972—SFP 2132 (A reading of John McGee's poem against a background of music and aerial photography of a T-38.)

Army films may be obtained through your nearest Army base. The Army offers a catalog and the name of your nearest base through:

Training Aids Management Agency
Training Material Support Division
Tobyhanna, PA 18466

Navy films may be obtained through the nearest appropriate Naval base, as follows:

If you live in:

Connecticut, Delaware, DC, Kentucky, Maine, Maryland, Massachusetts, New Hampshire, New Jersey, New York, Ohio, Pennsylvania, Puerto Rico, Rhode Island, Vermont, Virginia, West Virginia

Contact the Education and Training Support Center at:

US Naval Base
Norfolk, VA 23511

If you live in:

Alabama, Florida, Georgia, Mississippi, North Carolina, South Carolina, Tennessee

Contact the Education and Training Support Center at:

US Naval Base
Charleston, SC 29408

If you live in:

Colorado, Illinois, Indiana, Iowa, Kansas, Michigan, Minnesota, Nebraska, North Dakota, South Dakota, Wisconsin, Wyoming

Contact the Education and Training Support Center at:

US Naval Base
Great Lakes, IL 60088

. . . and the Blue Angels. (US Navy photo)

If you live in:

Alaska, Arizona, Arkansas, California, Hawaii, Idaho, Louisiana, Montana, Nevada, Oklahoma, Oregon, Texas, Utah, Washington

Contact the Education and Training Support Center at:

Film Library
Fleet Station P.O. Building
San Diego, CA 92132

Another source of films, this time on soaring, is the **National Soaring Museum, Harris Hill, Elmira, NY 14903.** They claim to have the most complete library of films on soaring in the world.

Aviation Museums

Fortunately there are many aviation museums in the United States, and many more throughout the world. This section lists some of the most interesting in the U.S., and it does so geographically.

Northeast

Washington, DC Let's start with the best and most dynamic—the **National Air and Space Museum (Smithsonian Institution)** in Washington, DC, located on Independence Avenue, between 4th and 7th Streets. The nearest airport is Washington National. This fabulous museum was opened in 1976 to rave reviews. And not only is it good, it was brought in under budget! The museum includes the original Wright Flyer, the original "Spirit

The Original Wright Flyer, spotted at the Air and Space Museum. (Tim Foster photo)

of St. Louis," an X-15, a Skylab, and many other historic aircraft. There is also a movie called *To Fly* (small admission charge), which you should try to see, and a planetarium. And there is a piece of moon rock you can actually touch. One visit won't be enough, so plan several. The National Air and Space Museum is open every day except Christmas Day from 10:00 AM to 5:30 PM, and it stays open to 9:00 PM in the summer. There is no admission charge. **Telephone (202) 628-4422.**

Hartford, CT Just outside the Bradley International Airport at Windsor Locks in Connecticut was the **Bradley Air Museum.** This was the fourth largest collection of airplanes in the U.S., featuring mostly military types, including a Douglas Globemaster, a Boeing B-17, a B-47 and many others. There was also an indoor collection, featuring some really interesting aircraft, including the only example of a hybrid Curtiss design—a T-tail (very fashionable), radial-engine-plus-jet-engine, tricycle-gear, carrier-borne fighter, the XF-15C. The museum was virtually wiped out by a tornado in 1979, and is now seeking financial support to rebuild and reopen. Tax deductible contributions may be sent to **Bradley Air Museum, Bradley International Airport, Windsor Locks, CT 06096.**

Philadelphia, PA If you drive through downtown Philadelphia, you can see an all-stainless-steel biplane-amphibian (it's a Budd BB-1) parked in front of a building, and around the corner, in back, a British Airways Boeing 707. When you see these you're at the **Franklin Institute,** which is at 20th Street and Benjamin Franklin Parkway. Inside there are a few other aircraft, including a 1911 Wright biplane. Admission is $2.00, and the museum is open year round from 10:00 AM to 5:00 PM. **Telephone (215) 448-1000.**

Newport News, VA The **U.S. Army Transportation Museum** is located at Fort Eustis, near I 64. Admission is free. and it is open daily all year, Mondays to Fridays, 8 AM to 5 PM, weekends and holidays 1 PM to 5 PM. This collection features mostly helicopters. **Telephone (804) 878-3603.**

Southeast

Ozark, AL The **U.S. Army Aviation Museum** is located at Fort Rucker, about 25 miles up US 231 from Dothan. It is open daily except Christmas Day from 10 AM to 5 PM (weekends and holidays 1 PM to 5 PM). Admission is free. Mostly spotter planes and helicopters are displayed here. **Telephone (205) 255-4507.**

Pensacola, FL The **Naval Aviation Museum** is located at Pensacola, Florida, at the U.S. Naval Air Station there. It is open daily all year from 9:00 AM to 5:00 PM, and admission is free. **Telephone (904) 452-3604.**

Florence, SC Right by the Florence, SC, airport, is an open-air collection of aircraft and missiles—the **Florence Air and Missile Museum.** This has a good collection of U.S. military aircraft of the last thirty years or so, including a Boeing B-47, a Douglas B-66, a Convair F 102, and others. Admission is $1.50, and the museum is open from 9:00 AM to dark all year round.

1909 Bleriot from Cole Palen's collection. (Tim Foster photo)

Central

Chicago, IL **The Museum of Science and Industry, East 57th Street and South Lakeshore Drive, Chicago** has a small collection of aircraft, including a Spitfire, a Curtiss Jenny, and a Boeing 40B biplane. The museum is open from 9:30 AM to 5:30 PM daily except Christmas Day. It closes at 4 PM on weekdays in the non-summer months. Admission is free. **Telephone (312) 684-1414.**

Dearborn, MI **The Henry Ford Museum and Greenfield Village,** located just south of US Route 12 between Southfield Road and Oakwood Boulevard in **Dearborn,** has many historic aircraft, including a 1909 Bleriot monoplane, the Fokker Trimotor that Byrd flew over the North Pole in 1926, the Ford Trimotor that he flew over the South Pole in 1929, a Pitcairn Autogyro, and many more. It is open daily all year from 9 AM to 6 PM (to 5 PM in the winter). Admission is $3 for adults. **Telephone (313) 271-1620.**

Omaha, NE At Offutt Air Force Base is the **Strategic Aerospace Museum,** which gives a history of the Strategic Air Command through its aircraft. Many "Bs" are there, including a B 17, B 29, B 36, B 52, B 58, plus some "Fs"—the F 86, F 101, F 102, and several others. An admission fee is charged. The Museum is open from 8 AM to 5 PM every day except Christmas, New Year's Day, and Easter Sunday. The telephone number is **(402) 292-2001.**

Dayton, OH The **U.S. Air Force Museum** is at Wright-Patterson Air Force Base, just off Route 4, east of Dayton, Ohio. Here you will see an enormous collection of such aircraft as the B 70 jet bomber, the old B 36 (the one with ten engines—six piston and four jet), along with many other rare U.S. and foreign military birds. The Air Force Museum is open daily from 9:00 AM to 5:00 PM (it opens at 10:00 AM on Saturdays and Sundays). Admission is free. Telephone **(513) 255-3284.**

The Air Force Museum, Dayton, Ohio. (Air Force Museum photo)

The XB 70 bomber is part of the Air Force Museum collection. (Air Force Museum photo)

Milwaukee, WI The **Experimental Aircraft Association** maintains a museum at Franklin, Wisconsin—a suburb of Milwaukee. It is open daily from 8:30 AM to 5:00 PM and on Sundays from 11:00 AM to 5:00 PM. It is located at 11311 West Forest Home Avenue, Franklin, WI 53132. The nearest airport is Hales Corners. **Telephone (414) 425-4860.**

Southwest

Harlingen, TX For flying activities, try to catch the **Confederate Air Force,** which is a collection of "Colonels" who have assembled many warbirds, made them airworthy, and fly them at the drop of a ten-gallon hat. Yes, they're located in Texas, at Rebel Field, Harlingen, near Brownsville. **Telephone (512) 425-1057.** They have a large static display, and they are open from 9:00 AM to 5:00 PM daily and from 1:00 PM to 6:00 PM on Sundays. Admission is $2.00 (children $1.00).

San Antonio, TX **The Museum of History and Traditions** features some interesting aircraft, including a twin Mustang (F 82), a Bell P 63 King Cobra, a Lockheed F 94 Starfire, and others. It is located at Lackland Air Force Base, admission is free, and it's open daily 7:30 AM to 4 PM (9 AM to 6 PM weekends and holidays). It's closed on Christmas Day. **Telephone (512) 671-3716.**

West

San Diego, CA **The San Diego Aero-Space Museum** is located at **1649 El Prado,** and is open daily from 10 AM to 4:30 PM. It displays replicas of the Spirit of St. Louis and the Wright Flyer, as well as a Zero, the Convair Sea Dart jet seaplane, and numerous others. Admission is free. **Telephone (714) 234-8291.**

Santa Ana, CA The **Movieland of the Air Museum** is well worth a visit, and is located on the Orange County Airport (SNA). It features many of the aircraft used by the late Paul Mantz and Frank Tallman when they flew for the movies. There is an admission charge, and the museum is open from 10 AM to 5 PM every day except Mondays. Telephone (714) 545-5021.

Canada

There is an excellent museum at Rockcliffe Airport, Ottawa, Ontario, and smaller collections at Ottawa International Airport and the Ottawa's **National Museum of Science and Technology. The National Aeronautical Collection** is open daily at Rockcliffe all year, from 9 AM to 9 PM, and is closed on Mondays in the winter. It has many historical Canadian aircraft, some of which are kept in flying condition. Admission is free. **Telephone (613) 998-3814.**

For more on historical aircraft, see the section on Tours (page 206) and the section on Antique Aircraft (page 96).

Books

There are two books of value to the museum goer. One is *Veteran and Vintage Aircraft*, by Leslie Hunt (Scribners) and the other is *Aircraft Museum Directory* (Quadrant Press).

Restaurants Made Out of Airplanes

Old railroad men used to buy old railroad cars and turn them into restaurants, hence the diner's characteristic look. The tradition appears to have been handed down to aviation, because there are a number of restaurants made out of old airplanes.

If you drive along Route 1 in Penndel, Pa., just east of Philadelphia, you'll see a Lockheed Super G Constellation mounted on a pedestal beside the road, bearing the sign of **Jim Flannery's Restaurant and Cocktail Lounge.** You can sit inside the Connie and have a cocktail, with absolutely no fear of flying.

Jim Flannery's restaurant, Penndel, PA. (Mary Foster photo)

Go to the **Don Q Inn,** at Dodgeville, Wis., (state highway 23) and you'll see a Boeing KC-97 Stratocruiser serving as a coffee shop. Dodgeville Municipal Airport is right next to the Inn. In Mexico City, there is a Canadair C-4 North Star (Canadian version of the DC-4), now called the **Avion Cafeteria,** located next to the Holiday Inn by the International Airport. Near Luqa Airport, in Malta, there is another Super

Lockheed Lodestar fast-food stand near Montreal. (Canadian Aviation photo)

Connie, now called the **Super Constellation Bar.** And on a highway north of Montreal, Canada, there is a Lockheed Lodestar now doing service as a fast-food stand. There are probably some more around, but that's a quick look.

The Beech plant at Wichita. (Beech photo)

Tours

Many aircraft plants and aviation operations offer guided tours of their facilities. Here is a listing of the main facilities on view:

Beech Aircraft offers plant tours twice a day, Mondays through Fridays, at 10 AM and 2 PM. No advance notice is required unless a large group of people is involved, in which case, contact the **Plant Tours Programs Department, Beech Aircraft Corp., Wichita, KS 67201, telephone (316) 681-8186.** The tour takes about an hour, and photography

Beech Barons on the line at Wichita. (Beech photo)

is not permitted. If you are driving, go to 9709 East Central, Wichita. You can fly in to Beech Airport, and the tower will tell you where to go if you say you're there for a plant tour.

Cessna Aircraft Company (5800 East Pawnee Road, Wichita, KS 67201, telephone (316) 685-9111) offers free organized tours of their plant every day, Mondays through Fridays at 10 AM and 1:30 PM. If you're flying, land at the Cessna Airport (CEA) and park on the main ramp, then go into the Delivery Center to arrange your tour. The tour takes about an hour. Groups of more than ten people should contact Cessna in advance with full details.

The General Aviation Division of Rockwell International has a plant tour every Friday, between 1 and 3 PM. You should arrange this in advance by contacting the **Public Relations Department, Rockwell International, 5001 North Rockwell, Bethany, OK 73008, telephone (405) 789-5000, extension 345.** There is a limit of

15 people. Driving, go to Wiley Post Airport, 50th and Rockwell. You can fly in there. Taxi to the Rockwell hangar and go into the Reception Area for the tour.

Piper Aircraft offers tours at all its plants—Lock Haven, PA, Santa Maria, CA, and Vero Beach and Lakeland, FL. **Lock Haven** tours are conducted twice a day, at 11 AM and 2:30 PM, every day, Monday through Friday. No advance notice is required unless you have a large group (telephone **(717) 748-6711,** and ask for **Customer Relations**). You can fly in to Lock Haven airport and park at the Piper Reception Center, following instructions you receive on unicom, 122.8 mHz. If you're driving, go to the main entrance of the Piper plant on Bald Eagle Street at the southwest side of the airport. For the **Santa Maria** plant, tours take place Tuesdays and Thursdays, schedule permitting. Advance notice is advised, by contacting the Manager of the **Delivery Center at (805) 922-8411, extension 237.** No more than six people can be accomodated in one tour. You can fly into the Santa Maria airport (SMX), parking in the transient area near gate 4, and go to the Piper Delivery Center, or you can drive to 3427 Skyway Drive, Santa Maria. The tour takes about two hours. For **Vero Beach,** tours take place twice a day, Monday through Friday, at 10 AM and 2 PM. No advance notice is required, and up to 30 people can be accommodated. Fly in to Vero Beach Municipal airport (VRB), park on the Piper ramp, and go into the Delivery Center for the tour. If you're driving, the tour starts in the Administration Building lobby, Piper Drive, Vero Beach. Call **(305) 567-4361** for further information. Ask for the Customer Relations representative. The tour takes about an hour. **Lakeland** tours are held daily, Monday through Friday, at 2 PM. You can fly in to Lakeland airport (LAL), park on the ramp in front of the Piper factory, and go into the main entrance of the factory. Or if you're driving, go to 3000 Medulla Road. Piper is located on the south side of the airport. The tour takes about an hour. If you have a large group (over 10 people), contact **Customer Relations** at **(813) 646-2911** for arrangements.

The author's Comanche at Piper's headquarters in Lock Haven, PA. (Chris Foster photo)

Mooney Aircraft Corporation offers tours at their Kerrville, Texas, plant. Contact them at **(512) 896-6000,** or **Box 72, Kerrville, TX 78028** for more details.

One place of great interest to the true aviation buff has to be **Edwards Air Force Base.** Tours are available there over a three-hour period between 9 AM and 4 PM daily, Monday through Friday, but three weeks advance notice is requested. Contact the **Office of Information, Edwards Air Force Base, CA 93523, telephone (805) 277-2345 or (805) 277-4127.** They won't take more than 50 people on a tour, and their schedule fills up fast, which is why advance notice is required. You can't fly in to Edwards unless you're in a military plane. A good nearby airport is Mojave—another fascinating place to visit, since a lot of civilian test flying goes on there—for example the Canadair Challenger, Burt Rutan's fleet of birds, and others. Mojave is about 15 miles away from Edwards. Driving to Edwards, you go to the Center Headquarters Building on Rosamond Boulevard. The security police at the gate will direct you. Photos may be taken on the tour, unless the escort advises otherwise.

Most FAA control towers, approach control, and ATC center facilities will admit pilots for familiarization tours, subject to workload and traffic. To arrange a visit, simply call the appropriate facility in advance and ask. No enormous amount of notice need be given unless you have a large group. For just one or two people, the best procedure is to call when ready and see if they'll take you. If you are a pilot, I strongly recommend that you visit FAA facilities of all types. You always learn something, and it helps the controllers to hear first-hand about the pilot's point of view.

NASA's **Lyndon B. Johnson Space Center** at Houston is open to the public every day of the week, except Christmas Day, between 9 AM and 4 PM. There is no admission charge. A collection of spacecraft and exhibits about America's space program can be viewed in the Visitor Information Center, Building 2 at

The rocket park at Johnson Space Center. The big rocket is a Saturn V. (NASA photo)

the LBJ Space Center, which is located about 25 miles south of Houston, Texas. Take routes 45S/75 to NASA Road 1, and go about 3 miles to the east. The visitor entrance is located on Second Street, north of NASA Road 1. Building 2 is located on Avenue D, just east of the visitors' parking lot. A self-guided tour is offered, and a guided tour can be arranged by reservation Mondays to Fridays by contacting the Center at **(713) 483-4321** on weekdays, or by writing to the **Public Services Branch, AP 4, NASA Johnson Space Center, Houston, TX 77058.** Give the number of people in your party and the date requested. Allow about two and a half hours for the tour. There is an auditorium, which shows NASA space films, and among the attractions to be seen is the Mission Simulation and Training Facility and the famous Mission Control Center (part of the guided tours only).

The **Kennedy Space Center at Cape Canaveral,** FL, is well worth a visit. Tours are offered every day except Christmas Day between 8 AM and dark. Go to the Visitors' Center off State Route 405, 6 miles east of US 1, south of Titusville, FL. Admission is free, and there is also a 50-mile bus tour, for which a modest charge is made. The bus tour takes about two hours and visits Mission Control, the various pads, and the huge Vehicle Assembly Building. The last bus tour starts two hours before sunset. Contact **NASA Tours, Box 21222, Kennedy Space Center, FL 32815, telephone (305) 269-3000, (800) 432-2153 within Florida.** Try to schedule your visit for a time when there is a spacecraft launching. Call the Center to find out when one is planned.

Another NASA facility offering tours is the **Langley Research Center,** in Virginia, three miles north of Hampton on Virginia Route 134. The Center is open to the public daily, Monday through Saturday from 8:30 AM to 4:30 PM and from noon to 4:30 PM on Sundays. Apart from some spacecraft, there is a wind-tunnel exhibit and a display featuring the Viking Mars program. **NASA Langley Visitor Center, Hampton, VA 23665, telephone (804) 827-2855.** Tours are available for groups by prior arrangement.

The Space Shuttle *Enterprise* on the launch pad at Kennedy Space Center. (NASA photo)

PART SIX

Etcetera

AIRLINES

AIRPORTS

ASSOCIATIONS

FEDERAL AVIATION AGENCY (FAA)

SCHOOLS

STATE AVIATION AGENCIES

PHONETIC ALPHABET AND MORSE CODE

KEY TO AIRSPEED TERMINOLOGY

ABBREVIATIONS

INTERNATIONAL CIVIL AIRCRAFT MARKINGS

The Dassault Falcon 50 trijet touches down.
(Falcon Jet photo)

Airlines

The airline picture in the United States has changed substantially in the last few years because of *deregulation*. Whereas in the past, such things as routes and fares were heavily regulated by the **Civil Aeronautics Board (CAB, Universal Building, 1825 Connecticut Avenue NW, Washington, DC 20428)**, they aren't now, and CAB won't be soon.

Here are the principal U.S. airlines, with the aircraft they either operate or have on order as of April 1979:

One of American's 155 727s. (American Airlines photo)

Air California
3636 Birch Street
Newport Beach, CA 92660

10 Boeing 737, 3 Lockheed Electra

Air Florida
3900 NW 79th Avenue
Miami, FL 33166

4 Boeing 737, 2 Convair 440, 5 Douglas DC-9, 4 Douglas DC 3

Air New England
Logan International Airport
East Boston, MA 02128

7 Fairchild FH-227, 10 DHC Twin Otter

Alaska Airlines
Box 68900
Sea-Tac International Airport
Seattle, WA 95158

12 Boeing 727

Aloha Airlines
Box 30028
Honolulu, HI 96820

10 Boeing 737

American Airlines
633 Third Avenue
New York, NY 10017

Braniff's "Big Orange." (Braniff International photo)

11 Boeing 747, 63 Boeing 707, 155 Boeing 727, 30 Boeing 767, 35 Douglas DC 10

Aspen Airways
Hangar 5
Stapleton International Airport
Denver, CO 80207

10 Convair 580

Braniff International
Braniff Boulevard
Dallas-Fort Worth Airport
TX 75261

11 Boeing 747, 117 Boeing 727, 16 Douglas DC 8

Capitol International Airways
Smyrna Airport
Nashville, TN 37167

12 Douglas DC 8

American's terminal at JFK. (American Airlines photo)

A Continental DC 10. (Continental photo)

Continental Airlines
Los Angeles International Airport
CA 90009

52 Boeing 727, 17 Douglas DC 10

Delta Airlines
Atlanta International Airport
GA 30320

132 Boeing 727, 20 Boeing 767, 23 Douglas DC 8, 49 Douglas DC 9, 38 Lockheed 1011

Eastern Airlines
Miami International Airport
FL 33148

23 Airbus A 300B, 133 Boeing 727, 84 Douglas DC 9, 34 Lockheed 1011

Evergreen International Airlines
Marana Air Park
Marana, AZ 85238

6 Boeing 727, 6 Convair 580, 6 Douglas DC 8, 3 Douglas DC 9, 6 Lockheed Electra

Frontier Airlines
8250 Smith Road
Denver, CO 80207

40 Boeing 737, 27 Convair 580, 3 DHC Twin Otter

Hawaiian Airlines
Box 30008
Honolulu, HI 96820

12 Douglas DC 9, 8 Lockheed Electra, 3 Shorts 330

Eastern's A 300B Airbus lands at Washington National for the first time. (Eastern Airlines photo)

Hughes Airwest
San Francisco International Airport
CA 94128

13 Boeing 727, 41 Douglas DC 9

National Airlines
Box 592055
Miami AMF, FL 33159
(merged with Pan American in 1979)

40 Boeing 727, 16 Douglas DC 10

Northwest Orient Airlines
Minneapolis-St. Paul International Airport
St. Paul, MN 55111

29 Boeing 747, 69 Boeing 727, 22 Douglas DC 10

Ozark Airlines
Lambert-St. Louis Municipal Airport,
MO 63145

2 Boeing 727, 40 Douglas DC 9, 14 Fairchild FH 227B

Pan Am's 747SP for trips like New York-Tokyo non-stop. (Pan American photo)

Pan Am's "WorldPort" at JFK. (Pan American photo)

Pacific Southwest Airlines
3225 North Harbor Drive
San Diego, CA 92101

34 Boeing 727, 12 Douglas DC 9

Pan American World Airways
Pan Am Building
New York, NY 10017
(see also National Airlines)

46 Boeing 747, 26 Boeing 707, 13 Boeing 727, 12 Lockheed 1011

Piedmont Airlines
Smith Reynolds Airport
Winston-Salem, NC 27102

6 Boeing 727, 28 Boeing 737, 17 YS-11A

Republic Airlines
7500 Airline Drive
Minneapolis, MN 55450

3 Boeing 727, 25 Convair 580, 68 Douglas DC 9, 8 Swearingen Metro

Southwest Airlines
1820 Regal Row
Dallas, TX 75235

1 Boeing 727, 18 Boeing 737

Texas International Airlines
Box 12788
Houston, TX 77017

3 Convair 600, 34 Douglas DC 9

Trans International Airlines
Oakland International Airport
Oakland, CA 94614

2 Boeing 747, 2 Douglas DC 8, 3 Douglas DC 10, 9 Lockheed Electra, 12 Lockheed Hercules

Trans World Airlines
605 Third Avenue
New York, NY 10016

14 Boeing 747, 91 Boeing 727, 85 Boeing 707, 14 Douglas DC 9, 26 Lockheed 1011

Republic Airlines DC 9. (Republic Airlines photo)

United Airlines DC 8-61. (United Airlines photo)

United Airlines
Box 66100
Chicago, IL 60666

30 Boeing 767, 18 Boeing 747, 184 Boeing 727, 59 Boeing 737, 70 Douglas DC 8, 42 Douglas DC 10

US Air
Washington National Airport
Washington, DC 20001

13 Boeing 727, 49 Douglas DC 9, 30 BAC 111, 8 Mohawk 298

US Air used to be Allegheny. (US Air photo)

Western Airlines
6060 Avion Drive
Los Angeles, CA 90045

5 Boeing 707, 9 Boeing 720, 44 Boeing 727, 21 Boeing 737, 13 Douglas DC 10

Wien Air Alaska
4100 International Airport Road
Anchorage, AK 99502

10 Boeing 737, 4 DHC Twin Otter, 4 Fairchild FH 227, 4 Fairchild F 27, 1 Grumman Mallard, 2 Shorts Skyvan

World Airways
Oakland International Airport
Oakland, CA 94614

3 Boeing 727, 1 Convair 340, 5 Douglas DC 8, 9 Douglas DC 10

Airports

Most dangerous airports

The International Federation of Airline Pilots Associations regularly rates international airline airports throughout the world from a safety point of view. According to IFALPA, the most dangerous airport in the United States is Los Angeles International (LAX), because of nighttime noise abatement procedures, which require aircraft to take off westbound over the Pacific Ocean and to land eastbound from over the sea. This causes head-to-head traffic patterns, which has earned LAX the only "black star" rating in the United States. IFALPA regularly updates these ratings, and recently removed Boston Logan (BOS) from its black star status due to improvements being made in noise-abatement procedures, runway grooving (for better braking action), and other safety changes.

Here are the "world's most dangerous (airline) airports," according to IFALPA. All of these hold the back star rating:

American Samoa (U.S.):
Pago Pago

Australia:
Kalgoorlie
Learmonth
Meekatharra

Colombia:
Bogota/Eldorado
Barranquilla/Cortissoz
Cali/Palmaseca
Cartagena/Crespo
Leticia/A. Vasquez Cobo
San Andres Island
Medellin, Olaya Herrera

Fiji:
Suva/Nausori

Greece:
Kerkira (Corfu)

Indonesia:
Ujung Pandang

Italy:
Rimini

Solomon Islands:
Honiara/Henderson

Tonga:
Tongagatapu/Fua' Amotu

United States:
Los Angeles International

Venezuela:
Caracas, Maiquetia
Maturin

Virgin Islands (U.S.):
St. Thomas H. S. Truman

Western Samoa:
Apia/Faleolo

Two other U.S. airports are rated by IFALPA, but with red stars (significant safety deficiencies). These are Anchorage International, Alaska, because it needs a crosswind roadway, and Honolulu, Hawaii, for various reasons. New York's JFK recently had its red star removed because of safety improvements.

This United Airlines DC 10 simulator can simulate any available airport on the visual display, and present all the features, by night or by day—a useful way to practice using hairy airports. (United Airlines photo)

Busiest U. S. airports

Each year the FAA reports on air traffic activity. Here are the busiest airports with FAA control towers during 1978:

Rank	Airport	State	Ident.	Movements
				(000)
1.	Chicago, O'Hare	IL	ORD	755.0
2.	Santa Ana, Orange County	CA	SNA	604.1
3.	Long Beach, Daugherty	CA	LGB	602.3
4.	Los Angeles, Van Nuys	CA	VNY	600.7
5.	Atlanta, Hartsfield	GA	ATL	544.0
6.	Los Angeles, International	CA	LAX	528.5
7.	Miami, Opa Locka	FL	OPF	482.6
8.	Denver, Stapleton	CO	DEN	468.6
9.	San Jose, Municipal	CA	SJC	458.6
10.	Oakland, International	CA	OAK	448.7
11.	Phoenix, Sky Harbor	AZ	PHX	420.1
12.	Seattle, Boeing	WA	BFI	409.6
13.	Torrance, Municipal	CA	TOA	405.8
14.	San Jose, Reid-Hillview	CA	RHV	402.9
15.	Dallas-Fort Worth Regional	TX	DFW	398.6
16.	Hayward	CA	HWD	395.6
17.	Denver, Arapahoe County	CO	APA	391.7
18.	Miami, Tamiami	FL	TMB	384.9
19.	Honolulu, International	HI	HNL	365.5
20.	Anchorage, Merrill	AK	MRI	365.2

You'll note that there are no New York airports in the top twenty. For the record, here's where they fall:

Rank	Airport	State	Ident.	Movements
21.	New York, La Guardia	NY	LGA	361.9
31.	New York, Kennedy	NY	JFK	339.5
47.	Teterboro	NJ	TEB	282.8
104.	Newark	NJ	EWR	205.7

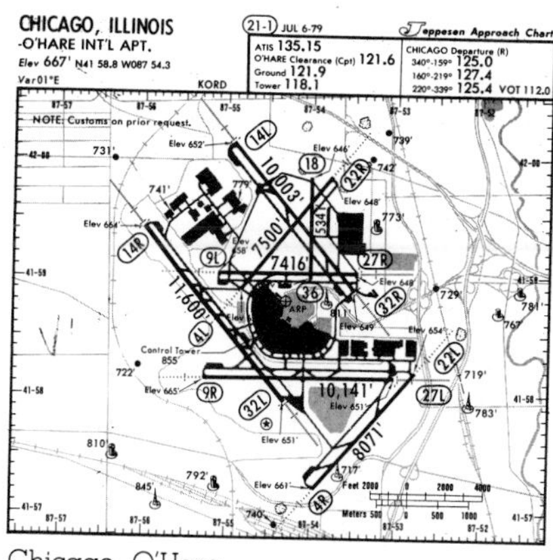

Chicago, O'Hare.

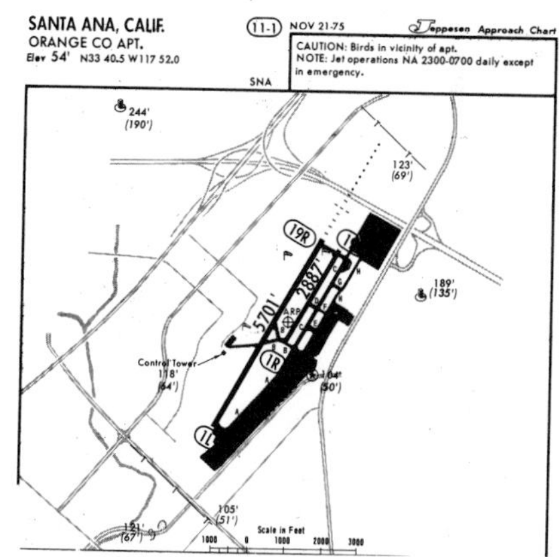

Santa Ana, Orange County.

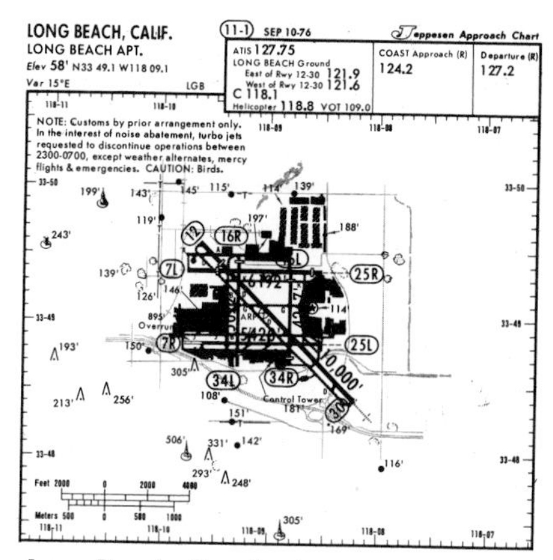

Long Beach, Daugherty.

Charts © 1980 Jeppesen-Sanderson. Reduced for illustration. Not to be used for navigation purposes. Used with permission.

Major Airport Codes—USA Alphabetical by City

City	State	Code
Abilene, Municipal	TX	ABI
Akron, Municipal*	OH	AKR
Akron/Canton	OH	CAK
Albany, County*	NY	ALB
Albany, Dougherty County	GA	ABY
Albuquerque, Intl.*	NM	ABQ
Allentown, -Bethlehem-Easton	PA	ABE
Alpena, Phelps Collins	MI	APN
Altoona, Peterson Memorial	PA	AQO
Amarillo, Intl.	TX	AMA
Anchorage, Intl.*	AK	ANC
Anchorage, Merrill	AK	MRI
Asheville	NC	AVL
Aspen, Pitkin County (Sardy)	CO	ASE
Athens, Municipal	GA	AHN
Atlanta, Charlie Brown County	GA	FTY
Atlanta, DeKalb Peachtree	GA	PDK
Atlanta, Hartsfield*	GA	ATL
Atlantic City, Municipal (Bader)	NJ	AIY
Atlantic City, NAFEC	NJ	ACY
Augusta, Bush	GA	AGS
Augusta, State	ME	AUG
Austin, Mueller Municipal	TX	AUS
Baker, Municipal	OR	BKE
Bakersfield, Meadows	CA	BFL
Baltimore, Balt-Wash Intl.*	MD	BAL
Bangor, Intl.*	ME	BGR
Banning, Municipal	CA	BNG
Bar Harbor	ME	BHB
Bartlesville, Phillips	OK	BVO
Baton Rouge, Ryan*	LA	BTR
Battle Creek, Kellogg*	MI	BTL
Baudette, Intl.*	MN	BDE
Beaumont, Jefferson County*	TX	BPT
Bedford, Hanscom*	MA	BED
Bellingham, Intl.*	WA	BLI
Benton Harbor, Ross	MI	BEH
Beverly, Municipal	MA	BVY
Billings, Logan Intl.*	MT	BIL
Binghamton, Broome County	NY	BGM
Birmingham, Municipal*	AL	BHM
Bismark, Municipal	ND	BIS

*Customs airport

City	State	Code
Block Island, State	RI	BID
Boise, Air Terminal (Gowen)	ID	BOI
Boston, Logan*	MA	BOS
Bowling Green, Warren County	KY	BWG
Bradford, -Regional	PA	BFD
Bridgeport, Sikorsky Memorial	CT	BDR
Brownsville, Intl.*	TX	BRO
Brownwood, Municipal	TX	BWD
Brunswick, McKinnon*	GA	SSI
Buffalo, Greater Buf Intl.*	NY	BUF
Burbank, Hollywood-	CA	BUR
Burlington, Intl.*	VT	BTV
Butte, Silver Bow County	MT	BTM
Cadillac, Wexford County	MI	CAD
Calexico, Intl.*	CA	CXL
Carbondale, Murphysboro	IL	MDH
Caribou, Municipal*	ME	CAR
Carlsbad, Cavern City Air Tmnl.	NM	CNM
Casper, Natrona County Intl.*	WY	CPR
Catskills, Sullivan County	NY	MSV
Cedar Rapids, Municipal	IA	CID
Champaign, University/Willard	IL	CMI
Charleston, AFB/Municipal*	SC	CHS
Charleston, Canawha County	WV	CRW
Charlotte, Douglas Municipal*	NC	CLT
Charlottesville, -Albemarle	VA	CHO
Chatanooga, Lovell*	TN	CHA
Cheyenne, Municipal	WY	CYS
Chicago, DuPage County	IL	DPA
Chicago, Meigs	IL	CGX
Chicago, Midway*	IL	MDW
Chicago, O'Hare*	IL	ORD
Chico, Municipal	CA	CIC
Cincinnati, Greater-*	OH	CVG
Cincinnati, Lunken	OH	LUK
Clarksburg, Benedum	WV	CKB
Cleveland, Cuyahoga County	OH	CGF
Cleveland, Hopkins Intl.*	OH	CLE
Cleveland, Lakefront	OH	BKL
Clinton, Municipal	IA	CWI
Colorado Springs, Peterson	CO	COS
Columbia, Metropolitan	SC	CAE
Columbia, Regional	MO	COU
Columbus, Metropolitan	GA	CSG

City	State	Code
Columbus, Ohio State University	OH	OSU
Columbus, Port Columbus Intl.*	OH	CMH
Concord, Buchanan	CA	CCR
Concord, Municipal	NH	CON
Corpus Christi, Intl.*	TX	CRP
Cut Bank*	MT	CTB
Dallas, Addison	TX	ADS
Dallas, Love Field*	TX	DAL
Dallas, Redbird	TX	RBD
Dallas/Ford Worth, Regional*	TX	DFW
Danville, Vermilion County	IL	DNV
Dayton, Cox*	OH	DAY
Dayton, General -South	OH	MGY
Dayton Beach, Regional	FL	DAB
Decatur	IL	DEC
Del Rio, Intl.*	TX	DRT
Denver, Arapahoe County	CO	APA
Denver, Jeffco	CO	BJC
Denver, Stapleton*	CO	DEN
Des Moines, Intl.*	IA	DSM
Detroit, City*	MI	DET
Detroit, Metropolitan Wayne Cty*	MI	DTW
Detroit, Willow Run	MI	YIP
Dodge City, Municipal	KS	DDC
Dothan	AL	DHN
Dubuque, Municipal	IA	DBQ
Duluth, Intl.*	MN	DLH
Eagle Pass, Municipal*	TX	EGP
East St. Louis, Bi-State Parks	IL	CPS
Eau Claire, Municipal	WI	EAU
El Paso, Intl.*	TX	ELP
Elkhart, Municipal	IN	EKM
Elkins, -Randolph County	WV	EKN
Elko, Municipal (Harris)	NV	EKO
Elmira, Chemung County	NY	ELM
Enid, Woodring Municipal	OK	WDG
Erie, Intl.*	PA	ERI
Escanaba, Delta County	MI	ESC
Eugene, Mahlon Sweet	OR	EUG
Eureka, Murray	CA	EKA
Evansville, Dress Regional	IN	EVV
Everett, Snohomish County*	WA	PAE
Fairbanks, Intl.*	AK	FAI
Fairbanks, Metro	AK	MTX
Fargo, Hector	ND	FAR
Farmingdale, Republic	NY	FRG
Flagstaff, Pulliam	AZ	FLG
Flint, Bishop	MI	FNT
Florence, City-County	SC	FLO
Fort Dodge, Municipal	IA	FOD
Fort Lauderdale, Executive*	FL	FXE
Fort Lauderdale, Hollywood Intl.*	FL	FLL
Fort Myers, Page	FL	FMY
Fort Pierce, Saint Lucie Cty	FL	FPR
Fort Smith, Municipal	AR	FSM
Fort Wayne, Baer	IN	FWA
Fort Worth, Meacham*	TX	FTW
Frankfort, Capital City	KY	FFT
Fresno, Air Terminal	CA	FAT
Fresno, Chandler Downtown	CA	FCH
Fullerton, Municipal	CA	FUL
Gainesville, Municipal	FL	GNV
Gaithersburg, Montgomery County	MD	GAI
Gallup, Clarke	NM	GUP
Galveston, Scholes*	TX	GLS
Garden City, Municipal	KS	GCK
Goodland, Municipal (Renner)	KS	GLD
Grand Canyon, National Park	AZ	GCN
Grand Forks, Intl.*	ND	GFK
Grand Island, Hall County Rgnl.	NE	GRI
Grand Junction, Walker	CO	GJT
Grand Rapids, Kent County*	MI	GRR
Great Falls, Intl.*	MT	GTF
Green Bay, Austin-Straubel	WI	GRB
Greenbrier, Valley	WV	LWB
Greensboro, High Point*	NC	GSO
Greenville, Municipal	MS	GLH
Greenville, Pitt	NC	PGV
Greenville/Spartanburg, Downtown	SC	GMU
Greenville/Spartanburg, Downtown	SC	SPA
Greenwood, County	SC	GRD
Gulfport, Municipal*	MS	GPT
Hagerstown, Regional	MD	HGR
Hana, Maui	HI	HNM
Harlingen, Industrial Air Park	TX	HRL
Harrisburg, Capital City	PA	HAR
Harrison, Boone County	AR	HRO
Hartford, Bradley Intl.*	CT	BDL

City	State	Code
Hartford, Brainard	CT	HFD
Havre, -City County*	MT	HVR
Hawthorne, Municipal	CA	HHR
Hayward, Air Terminal	CA	HWD
Helena	MT	HLN
Hilo, Lyman	HI	ITO
Hobbs, Lea County	NM	HOB
Honolulu, Intl.*	HI	HNL
Hot Springs, Ingalls	VA	HSP
Hot Springs, Memorial	AR	HOT
Houlton, Intl.*	ME	HUL
Houston, Hobby*	TX	HOU
Houston, Intercontinental*	TX	IAH
Huntington, Tri-State	WV	HTS
Huntsville, Madison Cty Jetport	AL	HSV
Huron, Regional	SD	HON
Hutchinson, Municipal (Stevens)	KS	HUT
Hyannis, Barnstable Municipal	MA	HYA
Idaho Falls, Fanning	ID	IDA
Indianapolis, Municipal*	IN	IND
International Falls, Falls Intl.*	MN	INL
Iowa City, Municipal	IA	IOW
Islip, MacArthur	NY	ISP
Ithaca, Tompkins County	NY	ITH
Jackson, County (Reynolds)	MI	JXN
Jackson, Hole	WY	JAC
Jackson, Thompson	MS	JAN
Jacksonville, Craig Municipal	FL	CRG
Jacksonville, Intl.*	FL	JAX
Jamestown, Chautaqua County	NY	JHW
Jefferson City, Memorial	MO	JEF
Johnstown, -Cambria County	PA	JST
Jonesboro, Municipal	AR	JBR
Joplin, Municipal	MO	JLN
Juneau, Intl.*	AK	JNU
Kalamazoo, Municipal	MI	AZO
Kansas City, Downtown*	MO	MKC
Kansas City, Intl.*	MO	MCI
Keene, Dillant-Hopkins	NH	EEN
Ketchikan, Intl.*	AK	KTN
Key West, Intl.*	FL	EYW
Kinston, Stallings	NC	ISO
Klamath Falls, Kingsley	OR	LMT
Knoxville, McGhee Tyson*	TN	TYS
Kodiak	AK	ADQ
Kodiak, Municipal	AK	KDK
Kokomo, Municipal	IN	OKK
LaCrosse, Municipal	WI	LSE
Laconia, Municipal	NH	LCI
Lafayette, Purdue University	IN	LAF
Lake Charles, Municipal*	LA	LCH
Lakeland, Municipal	FL	LAL
Lancaster	PA	LNS
Lansing, Capital City	MI	LAN
Laramie, Brees	WY	LAR
Laredo, Intl.*	TX	LRD
Las Cruces, -Crawford	NM	LRU
Las Vegas, McCarran*	NV	LAS
Las Vegas, North- Air Terminal	NV	VGT
Latrobe	PA	LBE
Lawrence, Municipal	MA	LWM
Lebanon, Regional	NH	LEB
Lewiston, Auburn- Municipal	ME	LEW
Lexington, Blue Grass	KY	LEX
Liberal, Municipal	KS	LBL
Lincoln, Municipal	NE	LNK
Linden	NJ	LDJ
Little Rock, Adams*	AR	LIT
Lock Haven, Piper Memorial	PA	LHV
London, -Corbin (MaGee)	KY	LOZ
Long Beach, Daugherty	CA	LGB
Los Angeles, Intl.*	CA	LAX
Los Angeles, Van Nuys	CA	VNY
Louisville, Bowman*	KY	LOU
Louisville, Standiford	KY	SDF
Lubbock, Regional*	TX	LBB
Lufkin, Angelina County	TX	LFK
Lynchburg, Municipal	VA	LYH
Macon, Wilson	GA	MCN
Madison, Dane Cty Rgnl. (Truax)	WI	MSN
Manchester, Municipal (Grenier)	NH	MHT
Mansfield, Lahm	OH	MFD
Marion, Williamson County	IL	MWA
Marquette, County*	MI	MQT
Martha's Vineyard	MA	MVY
Mason City, Municipal	IA	MCW
Massena, Richards*	NY	MSS
McAllen, Miller Intl.*	TX	MFE

City	State	Code
Medford, -Jackson County	OR	MFR
Melbourne, Kennedy Regional*	FL	MLB
Memphis, Intl.*	TN	MEM
Menominee, Twin County	MI	MNM
Merced, Municipal	CA	MCE
Miami, Intl.*	FL	MIA
Miami, New Tamiami	FL	TMB
Miami, Opa Locka*	FL	OPF
Middletown, Harrisburg-Olmsted*	PA	MDT
Midland, Air Park	TX	MDD
Midland, Regional	TX	MAF
Millville, Municipal	NJ	MIV
Milwaukee, Mitchell*	WI	MKE
Milwaukee, Timmerman	WI	MWC
Minneapolis, Anoka County-Janes	MN	ANE
Minneapolis, Crystal	MN	MIC
Minneapolis, Flying Cloud	MN	FCM
Minneapolis/St. Paul, Intl.*	MN	MSP
Minot, Intl.*	ND	MOT
Missoula, County (Johnson-Bell)	MT	MSO
Moab, Canyonlands	UT	CNY
Mobile, Bates*	AL	MOB
Modesto, City-County (Sham)	CA	MOD
Mojave	CA	MHV
Moline, Quad City	IL	MLI
Monroe, Municipal	LA	MLU
Monterey, Peninsula	CA	MRY
Montgomery, Dannelly*	AL	MGM
Morgantown, Municipal	WV	MGW
Mount Vernon, -Outland	IL	MVN
Muncie, Delaware County	IN	MIE
Muskegon, County	MI	MKG
Nantucket, Memorial	MA	ACK
Naples, Municipal	FL	APF
Nashville, Metropolitan*	TN	BNA
Natchez, Adams County (Anders)	MS	HEZ
New Bedford, Municipal	MA	EWB
New Bern, Simmons-Nott	NC	EWN
New Haven, Tweed-	CT	HVN
New Orleans, Intl.*	LA	MSY
New Orleans, Lakefront	LA	NEW
New York, Flushing	NY	FLU
New York, Kennedy Intl.*	NY	JFK
New York, La Guardia*	NY	LGA

City	State	Code
Newark, Intl.*	NJ	EWR
Newburgh, Stewart*	NY	SWF
Newport News, Patrick Henry*	VA	PHF
Newport, State	RI	NPT
Niagara Falls, Intl.*	NY	IAG
Nogales, Intl.*	AZ	OLS
Norfolk, Intl.*	VA	ORF
North Myrtle Beach, Grand Strand	SC	CRE
North Philadelphia	PA	PNE
North Platte, Lee Bird	NE	LBF
Norwood, Memorial	MA	OWD
Oakland, Intl.*	CA	OAK
Ocala, Municipal (Taylor)	FL	OCF
Ogdensburg, Intl.*	NY	OGS
Oklahoma City, Wiley Post	OK	PWA
Oklahoma City, Will Rogers World*	OK	OKC
Omaha, Eppy*	NE	OMA
Ontario, Intl.	CA	ONT
Orlando, Intl.*	FL	MCO
Oshkosh, Wittman	WI	OSH
Ottumwa, Industrial	IA	OTM
Owensboro, -Davies County	KY	OWB
Oxnard	CA	OXR
Paducah, Barkley	KY	PAH
Palm Springs, Municipal	CA	PSP
Panama City, Bay County (Fannin)	FL	PFN
Parkersburg, Wood County	WV	PKB
Paso Robles, Municipal	CA	PRB
Pawtucket, North Central State	RI	SFZ
Pellston, Emmett County	MI	PLN
Pembina, Municipal*	ND	PMB
Pendleton,Municipal	OR	PDT
Pensacola, Regional*	FL	PNS
Peoria, Greater-	IL	PIA
Philadelphia, Intl.*	PA	PHL
Philipsburg, Midstate	PA	PSB
Phoenix, Deer Valley Municipal	AZ	DVT
Phoenix, Sky Harbor Intl.*	AZ	PHX
Pierre, Municipal	SD	PIR
Pine Bluff, Grider	AR	PBF
Pittsburgh, Allegheny County	PA	AGC
Pittsburgh, Greater-*	PA	PIT
Pittsfield, Municipal	MA	PSF
Plattsburgh, Clinton County	NY	PLB

City	State	Code
Pontiac, Oakland-	MI	PTK
Port Huron, St. Clair County*	MI	PHN
Portland, Intl.*	OR	PDX
Portland, Intl. Jetport*	ME	PWM
Poughkeepsie, Dutchess County	NY	POU
Prescott, Municipal	AZ	PRC
Presque Isle, N. Maine Regional*	ME	PQI
Providence, Green State*	RI	PVD
Provincetown, Municipal	MA	PVC
Provo, Municipal	UT	PVU
Pueblo, Memorial	CO	PUB
Quincy, Municipal (Baldwin)	IL	UIN
Raleigh-Durham*	NC	RDU
Rapid City, Regional	SD	RAP
Reading, Municipal (Spaatz)	PA	RDG
Redding, Municipal	CA	RDD
Redmond, Roberts	OR	RDM
Reno, Intl.*	NV	RNO
Richmond, Byrd Intl.*	VA	RIC
Riverside, Municipal	CA	RAL
Roanoke, Municipal/Woodrum	VA	ROA
Rochester, Monroe County*	NY	ROC
Rochester, Municipal	MN	RST
Rockford, Greater-	IL	RFD
Rocky Mount, -Wilson	NC	RWI
Rocky Mount, Downtown	NC	RMT
Rosewell, Industrial Air Center	NM	ROW
Sacramento, Metropolitan*	CA	SMF
Saginaw, Tri City	MI	MBS
Salem, McNary	OR	SLE
Salina, Municipal	KS	SLN
Salisbury, Wicomico	MD	SBY
Salt Lake City, Intl.*	UT	SLC
San Angelo, Mathis	TX	SJT
San Antonio, Intl.*	TX	SAT
San Diego, Brown Municipal	CA	SDM
San Diego, Lindbergh*	CA	SAN
San Diego, Montgomery	CA	MYF
San Francisco, Intl.*	CA	SFO
San Jose, Municipal	CA	SJC
San Jose, Reid-Hillview	CA	RHV
San Luis Obispo, County	CA	SBP
Sandusky, Griffing-*	OH	SKY
Santa Ana, Orange County	CA	SNA

City	State	Code
Santa Barbara, Municipal	CA	SBA
Santa Fe, County Municipal	NM	SAF
Santa Maria, Public	CA	SMX
Santa Monica, Municipal	CA	SMO
Santa Rosa, Sonoma County	CA	STS
Sarasota/Bradenton	FL	SRQ
Sault Ste Marie, City-County	MI	SSM
Savannah, Municipal*	GA	SAV
Schenectady, County	NY	SCH
Seattle, Boeing Field*	WA	BFI
Seattle-Tacoma, Intl.*	WA	SEA
Sheboygan, County Memorial	WI	SBM
Shreveport Regional	LA	SHV
Shreveport, Downtown	LA	DTN
Sioux City, Municipal	IA	SUX
Sioux Falls, Joe Foss	SD	FSD
South Bend, Michiana Regional	IN	SBN
Southern Pines, Pinehurst-	NC	SOP
Spokane, Felts*	WA	SFF
Spokane, Intl.*	WA	GEG
Springfield, Capital	IL	SPI
St. Louis, Lambert/Intl.*	MO	STL
St. Louis, Spirit of St Louis	MO	SUS
St. Paul, Downtown (Holman)	MN	STP
St. Petersburg, Clearwater Intl.*	FL	PIE
Stockton, Metropolitan	CA	SCK
Syracuse, -Hancock Intl.*	NY	SYR
Tallahassee, Municipal	FL	TLH
Tampa, Intl.*	FL	TPA
Terre Haute, Hulman	IN	HUF
Teterboro, Municipal*	NJ	TEB
Texarkana, Municipal (Webb)	AR	TXK
Toledo, Express	OH	TOL
Toledo, Municipal	OH	TDZ
Topeka, Forbes	KS	FOE
Topeka, Municipal (Billard)	KS	TOP
Torrance, Municipal (Zamperini)	CA	TOA
Traverse City, Cherry Capital	MI	TVC
Trenton, Mercer County	NJ	TTN
Troutdale, Portland-	OR	TTD
Tucson, Intl.*	AZ	TUS
Tullahoma, William Northern	TN	THA
Tulsa, Intl.*	OK	TUL
Tulsa, Riverside	OK	RVS

City	State	Code
Tupelo, Lemons Municipal	MS	TUP
Tuscaloosa, Municipal	AL	TCL
Tyler, Pounds	TX	TYR
Utica, Oneida County	NY	UCA
Valparaiso, Porter County	IN	VPZ
Vero Beach, Municipal	FL	VRB
Visalia, Municipal	CA	VIS
Waco, -Madison Cooper	TX	ACT
Washington, Dulles*	DC	IAD
Washington, National	DC	DCA
Waterloo, Municipal	IA	ALO
Watertown, NY Intl.*	NY	ART
West Palm Beach, Intl.*	FL	PBI
Westerley, State	RI	WST
Westhampton Beach, Suffolk Cty	NY	FOK
White Plains, Westchester Cty*	NY	HPN
Wichita Falls, Municipal	TX	SPS
Wichita, Mid-Continent*	KS	ICT
Wildwood, Cape May County	NJ	WWD
Wilkes-Barre, Scranton*	PA	AVP
Williamsport, -Lycoming County	PA	IPT
Williston, Sloulin Intl.*	ND	ISN
Wilmington, Greater-*	DE	ILG
Wilmington, New Hanover County*	NC	ILM
Winslow, Municipal	AZ	INW
Winston Salem, Smith Reynolds*	NC	INT
Worcester, Municipal	MA	ORH

Piper Super Cub. (Piper photo)

Major Airport Codes
USA—Alphabetical by Code

Code	City	State
ABE	Allentown, -Bethlehem-Easton	PA
ABI	Abilene, Municipal	TX
ABQ	Albuquerque, Intl.*	NM
ABY	Albany, Dougherty County	GA
ACK	Nantucket, Memorial	MA
ACT	Waco, -Madison Cooper	TX
ACY	Atlantic City, NAFEC	NJ
ADQ	Kodiak	AK
ADS	Dallas, Addison	TX
AGC	Pittsburgh, Allegheny County	PA
AGS	Augusta, Bush	GA
AHN	Athens, Municipal	GA
AIY	Atlantic City, Municipal (Bader)	NJ
AKR	Akron, Municipal*	OH
ALB	Albany County*	NY
ALO	Waterloo, Municipal	IA
AMA	Amarillo, Intl.	TX
ANC	Anchorage, International*	AK
ANE	Minneapolis, Anoka County-Janes	MN
APA	Denver, Arapahoe County	CO
APF	Naples, Municipal	FL
APN	Alpena, Phelps Collins	MI
AQO	Altoona, Peterson Memorial	PA
ART	Watertown, NY Intl.*	NY
ASE	Aspen, Pitkin County (Sardy)	CO
ATL	Atlanta, Hartsfield*	GA
AUG	Augusta, State	ME
AUS	Austin, Mueller Municipal	TX
AVL	Asheville	NC
AVP	Wilkes-Barre, Scranton*	PA
AZO	Kalamazoo, Municipal	MI
BAL	Baltimore, Balt-Wash Intl.*	MD
BDE	Baudette, International*	MN
BDL	Hartford, Bradley Intl.*	CT
BDR	Bridgeport, Sikorsky Memorial	CT
BED	Bedford, Hanscom*	MA
BEH	Benton Harbor, Ross	MI
BFD	Bradford, -Regional	PA
BFI	Seattle, Boeing Field*	WA
BFL	Bakersfield, Meadows	CA
BGM	Binghamton, Broome County	NY

*Customs airport

Code	City	State
BGR	Bangor, Intl.*	ME
BHB	Bar Harbor	ME
BHM	Birmingham, Municipal*	AL
BID	Block Island, State	RI
BIL	Billings, Logan Intl.*	MT
BIS	Bismark, Municipal	ND
BJC	Denver, Jeffco	CO
BKE	Baker, Municipal	OR
BKL	Cleveland, Lakefront	OH
BLI	Bellingham, Intl.*	WA
BNA	Nashville, Metropolitan*	TN
BNG	Banning, Municipal	CA
BOI	Boise, Air Terminal (Gowen)	ID
BOS	Boston, Logan*	MA
BPT	Beaumont, Jefferson County*	TX
BRO	Brownsville, Intl.*	TX
BTL	Battle Creek, Kellogg*	MI
BTM	Butte, Silver Bow County	MT
BTR	Baton Rouge, Ryan*	LA
BTV	Burlington, Intl.*	VT
BUF	Buffalo, Greater Buf Intl.*	NY
BUR	Burbank, Hollywood-	CA
BVO	Bartlesville, Phillips	OK
BVY	Beverly, Municipal	MA
BWD	Brownwood, Municipal	TX
BWG	Bowling Green, Warren County	KY
CAD	Cadillac, Wexford County	MI
CAE	Columbia, Metropolitan	SC
CAK	Akron/Canton	OH
CAR	Caribou, Municipal*	ME
CCR	Concord, Buchanan	CA
CGF	Cleveland, Cuyahoga County	OH
CGX	Chicago, Meigs	IL
CHA	Chatanooga, Lovell*	TN
CHO	Charlottesville, -Albemarle	VA
CHS	Charleston, AFB/Municipal*	SC
CIC	Chico, Municipal	CA
CID	Cedar Rapids, Municipal	IA
CKB	Clarksburg, Benedum	WV
CLE	Cleveland, Hopkins, Intl.*	OH
CLT	Charlotte, Douglas Municipal*	NC
CMH	Columbus, Port Columbus Intl.*	OH
CMI	Champaign, University/Willard	IL
CNM	Carlsbad, Cavern City Air Tmnl.	NM

Code	City	State	Code	City	State
CNY	Maob, Canyonlands	UT	ERI	Erie, Intl.*	PA
CON	Concord, Municipal	NH	ESC	Escanaba, Delta County	MI
COS	Colorado Springs, Peterson	CO	EUG	Eugene, Mahlon Sweet	OR
COU	Columbia, Regional	MO	EVV	Evansville, Dress Regional	IN
CPR	Casper, Natrona County Intl.*	WY	EWB	New Bedford, Municipal	MA
CPS	East St. Louis, Bi-State Parks	IL	EWN	New Bern, Simmons-Nott	NC
CRE	North Myrtle Beach, Grand Strand	SC	EWR	Newark, Intl.*	NJ
CRG	Jacksonville, Craig Municipal	FL	EYW	Key West, Intl.*	FL
CRP	Corpus Christi, Intl.*	TX	FAI	Fairbanks, Intl.*	AK
CRW	Charleston, Canawha County	WV	FAR	Fargo, Hector	ND
CSG	Columbus, Metropolitan	GA	FAT	Fresno, Air Terminal	CA
CTB	Cut Bank*	MT	FCH	Fresno, Chandler Downtown	CA
CVG	Cincinnati, Greater*	OH	FCM	Minneapolis, Flying Cloud	MN
CWI	Clinton, Municipal	IA	FFT	Frankfort, Capital City	KY
CXL	Calexico, Intl.*	CA	FLG	Flagstaff, Pulliam	AZ
CYS	Cheyenne, Municipal	WY	FLL	Fort Lauderdale, Hollywood Intl.*	FL
DAB	Daytona Beach, Regional	FL	FLO	Florence, City-County	SC
DAL	Dallas, Love Field*	TX	FLU	New York, Flushing	NY
DAY	Dayton, Cox*	OH	FMY	Fort Myers, Page	FL
DBQ	Dubuque, Municipal	IA	FNT	Flint, Bishop	MI
DCA	Washington, National	DC	FOD	Fort Dodge, Municipal	IA
DDC	Dodge City, Municipal	KS	FOE	Topeka, Forbes	KS
DEC	Decatur	IL	FOK	Westhampton Beach, Suffolk County	NY
DEN	Denver, Stapleton*	CO	FPR	Fort Pierce, Saint Lucie Cty	FL
DET	Detroit, City*	MI	FRG	Farmingdale, Republic	NY
DFW	Dallas/Fort Worth, Regional*	TX	FSD	Sioux Falls, Joe Foss	SD
DHN	Dothan	AL	FSM	Fort Smith, Municipal	AR
DLH	Duluth, Intl.*	MN	FTW	Fort Worth, Meacham*	TX
DNV	Danville, Vermilion County	IL	FTY	Atlanta, Charlie Brown County	GA
DPA	Chicago, DuPage County	IL	FUL	Fullerton, Municipal	CA
DRT	Del Rio, Intl.*	TX	FWA	Fort Wayne, Baer	IN
DSM	Des Moines, Intl.*	IA	FXE	Fort Lauderdale, Executive*	FL
DTN	Shreveport, Downtown	LA	GAI	Gaithersburg, Montgomery County	MD
DTW	Detroit, Metropolitan Wayne Cty*	MI	GCK	Garden City, Municipal	KS
DVT	Phoenix, Deer Valley Municipal	AZ	GCN	Grand Canyon, National Park	AZ
EAU	Eau Claire, Municipal	WI	GEG	Spokane, Intl.*	WA
EEN	Keene, Dillant-Hopkins	NH	GFK	Grand Forks, Intl.*	ND
EGP	Eagle Pass, Municipal*	TX	GJT	Grand Junction, Walker	CO
EKA	Eureka, Murray	CA	GLD	Goodland, Municipal (Renner)	KS
EKM	Elkhart, Municipal	IN	GLH	Greenville, Municipal	MS
EKN	Elkins, -Randolph County	WV	GLS	Galveston, Scholes*	TX
EKO	Elko, Municipal (Harris)	NV	GMU	Greenville/Spartanburg, Dntn	SC
ELM	Elmira, Chemung County	NY	GNV	Gainesville, Municipal	FL
ELP	El Paso, Intl.*	TX	GPT	Gulfport, Municipal*	MS

Code	City	State	Code	City	State
GRB	Green Bay, Austin-Straubel	WI	JNU	Juneau, Intl.*	AK
GRD	Greenwood, County	SC	JST	Johnstown, -Cambria County	PA
GRI	Grand Island, Hall County Rgnl.	NE	JXN	Jackson, County (Reynolds)	MI
GRR	Grand Rapids, Kent County*	MI	KDK	Kodiak, Municipal	AK
GSO	Greensboro, High Point*	NC	KTN	Ketchikan, Intl.*	AK
GTF	Great Falls, Intl.*	MT	LAF	Lafayette, Purdue University	IN
GUP	Gallup, Clarke	NM	LAL	Lakeland, Municipal	FL
HAR	Harrisburg, Capital City	PA	LAN	Lansing, Capital City	MI
HEZ	Natchez, Adams County (Anders)	MS	LAS	Las Vegas, McCarran*	NV
HFD	Hartford, Brainard	CT	LAX	Los Angeles, Intl.*	CA
HGR	Hagerstown, Regional	MD	LBB	Lubbock, Regional*	TX
HHR	Hawthorne, Municipal	CA	LBE	Latrobe	PA
HLN	Helena	MT	LBF	North Platte, Lee Bird	NE
HNL	Honolulu, Intl.*	HI	LBL	Liberal, Municipal	KS
HNM	Hana, Maui	HI	LCH	Lake Charles, Municipal*	LA
HOB	Hobbs, Lea County	NM	LCI	Laconia, Municipal	NH
HON	Huron, Regional	SD	LDJ	Linden	NJ
HOT	Hot Springs, Memorial	AR	LEB	Lebanon, Regional	NH
HOU	Houston, Hobby*	TX	LEW	Lewiston, Auburn- Municipal	ME
HPN	White Plains, Westchester Cty.*	NY	LEX	Lexington, Blue Grass	KY
IAG	Niagara Falls, Intl.*	NY	LFK	Lufkin, Angelina County	TX
IAH	Houston, Intercontinental*	TX	LGA	New York, La Guardia*	NY
ICT	Wichita, Mid-Continent*	KS	LGB	Long Beach, Daugherty	CA
IDA	Idaho Falls, Fanning	ID	LHV	Lock Haven, Piper Memorial	PA
ILG	Wilmington, Greater-*	DE	LIT	Little Rock, Adams*	AR
ILM	Wilmington, New Hanover County*	NC	LMT	Klamath Falls, Kingsley	OR
IND	Indianapolis, Municipal*	IN	LNK	Lincoln, Municipal	NE
INL	International Falls, Falls Intl.*	MN	LNS	Lancaster	PA
INT	Winston Salem, Smith Reynolds*	NC	LOU	Louisville, Bowman*	KY
INW	Winslow, Municipal	AZ	LOZ	London, -Corbin (MaGee)	KY
IOW	Iowa City, Municipal	IA	LRD	Laredo, Intl.*	TX
IPT	Williamsport, -Lycoming County	PA	LRU	Las Cruces, -Crawford	NM
ISN	Williston, Sloulin Intl.*	ND	LUK	Cincinnati, Lunken	OH
ISO	Kinston, Stallings	NC	LWM	Lawrence, Municipal	MA
ISP	Islip, MacArthur	NY	MAF	Midland, Regional	TX
ITH	Ithaca, Tompkins County	NY	MBS	Saginaw, Tri City	MI
ITO	Hilo, Lyman	HI	MCE	Merced, Municipal	CA
JAN	Jackson, Thompson	MS	MCI	Kansas City, Intl.*	MO
JAX	Jacksonville, Intl.*	FL	MCN	Macon, Wilson	GA
JBR	Jonesboro, Municipal	AR	MCO	Orlando, Intl.*	FL
JEF	Jefferson City, Memorial	MO	MCW	Mason City, Municipal	IA
JFK	New York, Kennedy Intl.*	NY	MDD	Midland, Air Park	TX
JHW	Jamestown, Chautauqua County	NY	MDH	Carbondale, Murphysboro	IL
JLN	Joplin, Municipal	MO	MDT	Middletown, Harrisburg-Olmsted*	PA

Code	City	State	Code	City	State
MDW	Chicago, Midway*	IL	ONT	Ontario, Intl.*	CA
MEM	Memphis, Intl.*	TN	OPF	Miami, Opa Locka*	FL
MFD	Mansfield, Lahm	OH	ORD	Chicago, O'Hare*	IL
MFE	McAllen, Miller Intl.*	TX	ORH	Worcester, Municipal	MA
MFR	Medford, -Jackson County	OR	OSU	Columbus, Ohio State University	OH
MGM	Montgomery, Dannelly*	AL	OTM	Ottumwa, Industrial	IA
MGY	Dayton, General -South	OH	OWB	Owensboro, -Davies County	KY
MHT	Manchester, Municipal (Grenier)	NH	OWD	Norwood, Memorial	MA
MHV	Mojave	CA	OXR	Oxnard	CA
MIA	Miami, Intl.*	FL	PAH	Paducah, Barkley	KY
MIC	Minneapolis, Crystal	MN	PBF	Pine Bluff, Grider	AR
MIE	Muncie, Delaware County	IN	PBI	West Palm Beach, Intl.*	FL
MIV	Millville, Municipal	NJ	PDK	Atlanta, DeKalb Peachtree	GA
MKC	Kansas City, Downtown*	MO	PDT	Pendleton, Municipal	OR
MKG	Muskegon, County	MI	PDX	Portland, Intl.*	OR
MLB	Melbourne, Kennedy Regional*	FL	PFN	Panama City, Bay County (Fannin)	FL
MLI	Moline, Quad City	IL	PGV	Greenville, Pitt	NC
MLU	Monroe, Municipal	LA	PHL	Philadelpia, Intl.*	PA
MNM	Menominee, Twin County	MI	PHN	Port Huron, St. Clair County*	MI
MOB	Mobile, Bates*	AL	PHX	Phoenix, Sky Harbor Intl.*	AZ
MOD	Modesto, City-County (Sham)	CA	PIA	Peoria, Greater-	IL
MOT	Minot, Intl.*	ND	PIE	St. Petersburg, Clearwater Intl.*	FL
MQT	Marquette, County*	MI	PIR	Pierre, Municipal	SD
MRI	Anchorage, Merrill	AK	PIT	Pittsburgh, Greater-*	PA
MRY	Monterey, Peninsula	CA	PLB	Plattsburgh, Clinton County	NY
MSO	Missoula, County (Johnson-Bell)	MT	PLN	Pellston, Emmett County	MI
MSP	Minneapolis/St. Paul, Intl.*	MN	PMB	Pembina, Municipal*	ND
MSS	Massena, Richards*	NY	PNE	North Philadelphia	PA
MSV	Catskills, Sullivan County	NY	PNS	Pensacola, Regional*	FL
MSY	New Orleans, Intl.*	LA	POU	Poughkeepsie, Dutchess County	NY
MTX	Fairbanks, Metro	AK	PQI	Presque Isle, N. Maine Regional*	ME
MVN	Mount Vernon, -Outland	IL	PRB	Paso Robles, Municipal	CA
MVY	Martha's Vineyard	MA	PRC	Prescott, Municipal	AZ
MWA	Marion, Williamson County	IL	PSB	Philipsburg, Midstate	PA
MYF	San Diego, Montgomery	CA	PSF	Pittsfield, Municipal	MA
NEW	New Orleans, Lakefront	LA	PSP	Palm Springs, Municipal	CA
NPT	Newport, State	RI	PTK	Pontiac, Oakland-	MI
OAK	Oakland, Intl.*	CA	PUB	Pueblo, Memorial	CO
OCF	Ocala, Municipal (Taylor)	FL	PVC	Provincetown, Municipal	MA
OGS	Ogdensburg, Intl.*	NY	PVD	Providence, Green State*	RI
OKC	Oklahoma City, Will Rogers World*	OK	PVU	Provo, Municipal	UT
OKK	Kokomo, Municipal	IN	PWA	Oklahoma City, Wiley Post	OK
OLS	Nogales, Intl.*	AZ	PWM	Portland, Intl Jetport*	ME
OMA	Omaha, Eppy*	NE	RAL	Riverside, Municipal	CA

Code	City	State
RAP	Rapid City, Regional	SD
RBD	Dallas, Redbird	TX
RDD	Redding, Municipal	CA
RDG	Reading, Municipal (Spaatz)	PA
RDM	Redmond, Roberts	OR
RDU	Raleigh-Durham*	NC
RFD	Rockford, Greater-	IL
RHV	San Jose, Reid-Hillview	CA
RMT	Rocky Mount, Downtown	NC
RNO	Reno, Intl.*	NV
ROC	Rochester, Monroe County*	NY
ROW	Roswell, Industrial Air Center	NM
RST	Rochester, Municipal	MN
RVS	Tulsa, Riverside	OK
RWI	Rocky Mount, -Wilson	NC
SAF	Santa Fe, County Municipal	NM
SAN	San Diego, Lindbergh*	CA
SAT	San Antonio, Intl.*	TX
SAV	Savannah, Municipal*	GA
SBA	Santa Barbara, Municipal	CA
SBN	South Bend, Michiana Regional	IN
SBP	San Luis Obispo, County	CA
SBY	Salisbury, Wicomico	MD
SCH	Schenectady, County	NY
SCK	Stockton, Metropolitan	CA
SDF	Louisville, Standiford	KY
SDM	San Diego, Brown Municipal	CA
SFO	San Francisco, Intl.*	CA
SFZ	Pawtucket, North Central State	RI
SHV	Shreveport Regional	LA
SJC	San Jose, Municipal	CA
SJT	San Angelo, Mathis	TX
SKY	Sandusky, Griffing-*	OH
SLC	Salt Lake City, Intl.*	UT
SLE	Salem, McNary	OR
SLN	Salina, Municipal	KS
SMF	Sacramento, Metropolitan*	CA
SMO	Santa Monica, Municipal	CA
SMX	Santa Maria, Public	CA
SNA	Santa Ana, Orange County	CA
SOP	Southern Pines, Pinehurst	NC
SPA	Greenville/Spartanburg, Dntn.	SC
SPI	Springfield, Capital	IL

Code	City	State
SPS	Wichita Falls, Municipal	TX
SRQ	Sarasota/Bradenton	FL
SSI	Brunswick, McKinnon*	GA
SSM	Sault Ste Marie, City-County	MI
STL	St. Louis, Lambert/Intl.*	MO
STP	St. Paul, Downtown (Holman)	MN
STS	Santa Rosa, Sonoma County	CA
SUS	St. Louis, Spirit of St. Louis	MO
SUX	Sioux City, Municipal	IA
SWF	Newburgh, Stewart*	NY
SYR	Syracuse, -Hancock, Intl.*	NY
TCL	Tuscaloosa, Municipal	AL
TDZ	Toledo, Municipal	OH
TEB	Teterboro, Municipal*	NJ
THA	Tullahoma, William Northern	TN
TLH	Tallahassee, Municipal	FL
TMB	Miami, New Tamiami	FL
TOA	Torrance Municipal (Zamperini)	CA
TOL	Toledo, Express	OH
TOP	Topeka, Municipal (Billard)	KS
TPA	Tampa, Intl.*	FL
TTD	Troutdale, Portland-	OR
TTN	Trenton, Mercer County	NJ
TUL	Tulsa, Intl.*	OK
TUP	Tupelo, Lemons Municipal	MS
TUS	Tucson, Intl.*	AZ
TVC	Traverse City, Cherry Capital	MI
TXK	Texarkana, Municipal (Webb)	AR
TYR	Tyler, Pounds	TX
TYS	Knoxville, McGhee Tyson*	TN
UCA	Utica, Oneida County	NY
UIN	Quincy, Municipal (Baldwin)	IL
VGT	Las Vegas, North- Air Terminal	NV
VIS	Visalia, Municipal	CA
VNY	Los Angeles, Van Nuys	CA
VPZ	Valparaiso, Porter County	IN
VRB	Vero Beach, Municipal	FL
WDG	Enid, Woodring Municipal	OK
WST	Westerley, State	RI
WWD	Wildwood, Cape May County	NJ
YIP	Detroit, Willow Run	MI

Associations

Pilots are a clannish bunch, so they tend to form into groups at the flash of a strobelight. There are groups representing pilots and aircraft owners (Aircraft Owners and Pilots Association) groups representing specific aircraft owners (American Navion Society, Cessna 190-195 Owners Association, International Comanche Society), groups representing women pilots (The Ninety-Nines Inc.) and all kinds of other special interest groups (Experimental Aircraft Association, Flying Chiropractors Association, National Association of Priest Pilots), to say nothing of straight unions (Airline Pilots Association).

Some groups just send out a newsletter, while others organize massive annual events, such as the Experimental Aircraft Association annual fly-in at Oshkosh, Wisconsin, or the AOPA Plantation Party at Las Vegas, Nevada, or Miami, Florida. The EAA Fly-In makes Oshkosh the busiest airport in the world for the duration, with over 9,000 movements a day. The AOPA has over 230,000 members, so it is a force to be reckoned with, although its fly-in is less hectic.

Whatever your particular relationships with airplanes may be, somebody has formed an association of like-minded people, and if they haven't, go to it!

This listing shows most of the organizations I've been able to find, along with their *raison d'etre* if this is not obvious from the name. If you are interested in joining one of these, or you want to know more, write to the address given, which was accurate at the time of publication.

Aircraft Owners and Pilots Association
Box 5800
Washington, DC 20014
(301) 654-0500

American Bonanza Society
Box 3749
Reading, PA 19605
(215) 372-6967

American Navion Society
Municipal Airport
Banning, CA 92220
(714) 849-2213

Antique Airplane Association
Box H
Ottumwa, IA 52501
(515) 938-2773

Aviation Distributors and Manufacturers Association
1900 Arch Street
Philadelphia, PA 19103
(215) 564-3484

Aviation Historical Foundation
Suite 782
National Press Building
Washington, DC 20045
(202) 347-3963

Aviation Maintenance Foundation
Box 739
Basin, WY 82410
(307) 568-2466

Aviation/Space Writers Association
c/o William F. Kaiser
Cliffwood Road
Chester, NJ 07930
(201) 879-5667

Cessna 120/140 Association
Box 92
Richardson, TX 75080
(214) 234-2064

Cessna Skyhawk/Skylane Association
3 Lafayette Court
Camden, SC 29020
(803) 432-3586

The joy of Associating. (Beech photo)

Cessna 190/195 Owners Association
Box 952
Sioux Falls, ND 57101
(605) 338-1310

The Continental Luscombe Association
5736 Esmar Road
Ceres, CA 95307

Ercoupe Owners Club
Box 15058
Durham, NC 27704
(919) 477-2193

Experimental Aircraft Association
Box 229
Hales Corners, WI 53130
(414) 425-4860

Flying Architects Association
c/o H. B. Southern
571 East Hazlewood Avenue
Rahway, NJ 07065
(201) 388-5298

Flying Chiropractors Association
215 Belmont Street
Johnstown, PA 15904
(814) 266-3314

Flying Dentists Association
5820 Wilshire Boulevard
Los Angeles, CA 90036
(213) 937-5514

Flying Engineers International
Box 387
Winnebago, IL 61088
(813) 335-2660

Flying Funeral Directors of America
811 Grant Street
Akron, OH 44311
(216) 253-8121

Flying Optometrist Association of America
311 North Spruce
Searcy, AR 72143
(501) 268-3577

Flying Physicians Association
801 Green Bay Road
Lake Bluff, IL 60044
(312) 234-6330

Flying Psychologists
190 N Oakland Avenue
Pasadena, CA 91101
(213) 449-2182

Flying Veterinarians Association
10519 Reading Road
Cincinnati, OH 45241
(513) 563-0410

Future Airline Pilots of America
Suite 727
1515 E Tropicana Avenue
Las Vegas, NV 89109
(702) 739-7043
(800) 634-6166

General Aviation Manufacturers Association
Suite 517
1025 Connecticut Avenue NW
Washington, DC 20036
(202) 296-8848

The Helicopter Association of America
Suite 610
1156 15th Street NW
Washington, DC 20005
(202) 466-2420

International 180/185 Club
4539 N 49th Avenue
Phoenix, AZ 85031

International Aerobatic Club (Division of EAA)
Box 229
Hales Corners, WI 53130
(414) 425-4860

International Cessna 170 Association
Box 460
Camp Verde, AZ 86322

International Citabria Club
Box 29
White Lake, NY 12786
(914) 583-4070

International Comanche Society
Box 547
Grant, NB 69140
(308) 352-4495

International Flying Bankers Association
Box 11187
Columbia, SC 29211
(803) 252-5646

International Flying Farmers
Box 9124
Wichita, KS 67277
(316) 943-4234

International Swift Association
Box 644
Athens, TN 37303
(615) 745-9547

International Twin Engine Society
Suite 126
1499 Bayshore Highway
Burlingame, CA 94010
(415) 697-1373

Lawyer-Pilots Bar Association
2098 First National Bank Tower
Portland, OR 97201
(503) 224-3113

Mooney Aircraft Pilots Association
314 Stardust Drive
San Antonio, TX 78228
(512) 434-5959

National Aeronautic Association
Suite 430, 821 15th Avenue NW
Washington, DC 20005
(202) 347-2808

National Agricultural Aviation Association
Suite 459
National Press Building
Washington, DC 20045
(202) 638-0542

National Association of Flight Instructors
Box 20204
Columbus, OH 43220
(614) 459-0204

National Association of Priest Pilots
5157 South California Avenue
Chicago, IL 60632
(312) 434-3613

National Business Aircraft Association
One Farragut Square South
Washington, DC 20006
(202) 783-9000

National Intercollegiate Flying Association
Parks College
St. Louis University
Cahokia, IL 62206
(618) 337-7500

National Police Pilots Association
Box 54
Shenorock, NY 10587
(914) 682-5324

National Waco Club
2650 West Alexander Bell Road
Dayton, OH 45459
(513) 435-9725

The Ninety-Nines, Inc. (Women Pilots Association)
Box 59965
Oklahoma City, OK 73159
(405) 685-7969

Organized Flying Adjusters
Box 6391
Corpus Christi, TX 78411
(512) 853-0331

Pilots International Association
Suite 500
400 S. County Road 18
Minneapolis, MN 55426
(612) 546-4075

Pilot's Lobby
Box 1515
Washington, DC 20013
(202) 546-5150

Popular Rotorcraft Association
Box 570
Stanton, CA 90680
(714) 898-6555

Professional Aviation Maintenance Association
Box 12449
Pittsburgh, PA 15231
(412) 433-4825

Seaplane Pilots Association
Box 30091
Washington, DC 20014
(301) 951-3895

The Soaring Society of America
Box 66071
Los Angeles, CA 90066
(213) 390-4448

Taildragger Pilots Association
Box 161079
Memphis, TN 38116
(901) 345-7367

Tri Pacer Owners Club
c/o Robert H. Fuller
353 Nassau Street
Princeton, NJ 08540
(609) 921-0099

US Air Racing Association
16644 Roscoe Boulevard
Van Nuys, CA 91406
(213) 988-1751

US Parachute Association
Suite 444, 806 15th Street NW
Washington, DC 20005
(202) 737-0773

Wheelchair Pilots Association
12623 111th Lane North
Largo, FL 33540
(813) 581-5461

The Whirly- Girls, Inc.
Suite 700
1725 DeSales Street NW
Washington, DC 20036
(202) 347-2315

A reason to join the Seaplane Pilot's Association? (Edo-Aire photo)

The FAA does not regulate hang-glider operations. (Volmer photo)

Federal Aviation Agency (FAA)

The FAA is the chief regulatory agency affecting aviation in the U.S. It is a unit of the Department of Transportation (DOT). The **DOT** is at **400 Seventh Street SW, Washington, DC 20591, telephone (202) 655-4000. FAA HQ** is at **800 Independence Avenue SW, Washington, DC 20591, telephone (202) 426-4000.** The FAA Office of Public Affairs is located at this address.

Much of the FAA activity affecting aviators takes place at the **FAA Aeronautical Center, Box 25082, Oklahoma City, OK 73125, telephone (405) 686-2011.** This is located at Will Rogers Field, the main airport in Oklahoma City. Three important facilities dealing directly with the public are located there—the **Aircraft Registration Branch, telephone (405) 686-2116,** which deals with registering aircraft, recording titles, in issuing custom "N" numbers, and so on; the **Airmen Certification Branch, telephone (405) 686-2261,** which is where you write if you want to change the address on your pilot's license, or you want to replace a lost one; and the **Aeromedical Certification Branch, telephone (405) 686-4821,** which deals with medical certification matters.

The FAA maintains a testing operation in Atlantic City. It is the **National Aviation Facilities Experimental Center (NAFEC), Atlantic City, NJ 08405, telephone (609) 641-8200.**

The FAA divides its attention over 11 regions:

Region	Location	Governing area
Alaskan	Anchorage	AK, Aleutians
Central	Kansas City	IA, KS, MO, NE
Eastern	New York City	DE, DC, MD, NJ, NY, PA, VA, WV
Great Lakes	Chicago	IL, IN, MN, MI, OH, WI
New England	Boston	CT, ME, MA, NH, RI, VT
Northwest	Seattle	ID, OR, WA
Pacific	Honolulu	HI, Western Pacific
Rocky Mountain	Denver	CO, MT, ND, SD, UT, WY
Southern	Atlanta	AL, FL, GA, MS, NC, SC, TN, KY, PR, VI, CZ
Southwest	Fort Worth	AR, LA, NM, OK, TX
Western	Los Angeles	AZ, CA, NV

Within each Region are several General Aviation District Offices (GADOs), as follows:

Alaskan Region:

Anchorage, AK	(907) 272-1324
Fairbanks, AK	(907) 452-1276
Juneau, AK	(907) 789-0231

Central Region:

Des Moines, IA	(515) 284-4094
Kansas City, KS	(913) 281-3491
Wichita, KS	(316) 943-3244
St. Louis, MO	(314) 425-7102
Lincoln, NE	(402) 471-5485

Eastern Region:

Washington, DC	(202) 628-1555
Baltimore, MD	(301) 761-2610
Teterboro, NJ	(201) 288-1745
Albany, NY	(518) 869-8482
Farmingdale, NY	(516) 691-3100
Rochester, NY	(716) 263-5880
Allentown, PA	(215) 264-2888
New Cumberland, PA	(717) 782-4528
Philadelphia, PA	(215) 597-9708
Pittsburgh, PA	(412) 461-7800
Richmond, VA	(804) 222-7494
Charleston, WV	(304) 343-4689

Great Lakes Region:

Chicago, IL	(312) 584-4490
Springfield, IL	(217) 525-4238
Indianapolis, IN	(317) 247-2491
South Bend, IN	(219) 232-5843
Detroit, MI	(313) 485-2250
Grand Rapids, MI	(616) 456-6427
Minneapolis, MN	(612) 725-3341
Cincinnati, OH	(513) 684-2183
Cleveland, OH	(216) 267-0220
Columbus, OH	(614) 469-7476
Milwaukee, WI	(414) 747-5531

New England Region:

Portland, ME	(207) 744-4484
Boston, MA	(617) 762-2436
Westfield, MA	(413) 568-3121

Northwest Region:

Boise, ID	(208) 384-1238
Eugene, OR	(503) 688-9721
Portland, OR	(503) 221-2104
Seattle, WA	(206) 767-2747
Spokane, WA	(509) 456-4618

Pacific Region:

Honolulu, HI	(808) 847-0615

Rocky Mountain Region:

Denver, CO	(303) 466-7326
Billings, MT	(406) 245-6179
Helena, MT	(406) 449-5270
Fargo, ND	(701) 232-8949
Rapid City, SD	(605) 343-2403
Salt Lake City, UT	(801) 524-4247
Casper, WY	(307) 234-8959

Southern Region:

Birmingham, AL	(205) 254-1393
Jacksonville, FL	(904) 641-7311
Miami, FL	(305) 681-7431
St. Petersburg, FL	(813) 531-1434
Atlanta, GA	(404) 221-6481
Louisville, KY	(502) 582-6116
Jackson, MS	(601) 969-4633
Charlotte, NC	(704) 392-3214
Raleigh, NC	(919) 755-4240
West Columbia, SC	(803) 765-5931
Memphis, TN	(901) 345-0600
Nashville, TN	(615) 251-5661
San Juan, PR	(809) 791-5050

Southwest Region:

Little Rock, AR	(501) 372-3437
New Orleans, LA	(504) 241-2506
Shreveport, LA	(318) 226-5379
Albuquerque, NM	(505) 247-0156
Oklahoma City, OK	(405) 789-5220
Tulsa, OK	(918) 835-7619
Midland, TX	(915) 563-0802
Lafayette, LA	(318) 234-2321
Dallas, TX	(214) 357-0142
Corpus Christi, TX	(512) 884-9331
El Paso, TX	(915) 778-6389
Houston, TX	(713) 643-6504
Lubbock, TX	(806) 762-0335
San Antonio, TX	(512) 824-9535

Western Region:

Phoenix, AZ	(602) 261-4763
Fresno, CA	(209) 487-5306
Oakland, CA	(415) 273-7155
Ontario, CA	(714) 984-2411
Sacramento, CA	(916) 440-3169
San Diego, CA	(714) 293-5280
San Jose, CA	(408) 275-7681
Santa Monica, CA	(213) 391-6701
Van Nuys, CA	(213) 997-3191
Las Vegas, NV	(702) 736-0666
Reno, NV	(702) 784-5321
Long Beach, CA	(213) 426-7135

Flight Service Stations (FSS)

There are dozens of Flight Service Stations throughout the United States. These are staffed by FAA specialists who provide pilots with weather and facility briefings in advance of and during flights. You file your flight plan through the FSS. To locate the nearest one, look in the local phone book under "U.S. Government—Federal Aviation Agency." You may find a variety of listings, such as for Pilots' Automatic Telephone Weather Answering Service (PATWAS)—a continuous recording of local and regional weather; Flight Plan Fast File—a means of filing a flight plan (VFR or IFR) directly onto a recording machine; and possibly some toll-free (800) numbers for use in outlying areas. In some areas the Airport Terminal Information Service (ATIS) recording is available by phone, as well. Many airports have a direct phone line to the FSS. You can also reach the FSS on your aircraft radio. The frequencies are printed on the flight charts.

Enroute Flight Advisory Service (EFAS)

EFAS is a recent innovation that provides pilots with a standard radio frequency (122.0 mHz) to talk to a specialist at the FSS about weather only. It is not for filing flight plans or other purposes. These facilities are located all over the country. The trouble is the frequency—which is the same everywhere—has become very congested. When you're in the air, you could be talking to Pittsburgh and be cut out by some guy in a jet talking to Teterboro, hundreds of miles away.

Federal Aviation Regulations

FARs are issued by the Federal Aviation Agency (FAA), a unit of the Department of Transportation (DOT). FARs are divided into several parts. The most pertinent of these are:

FAR Part	*Subject*
1	Definitions and abbreviations
13	Enforcement procedures
21	Certification: products and parts
23	Airworthiness standards: normal, utility, and acrobatic category aircraft
25	Airworthiness standards: transport category aircraft
27	Airworthiness standards: normal category rotorcraft
29	Airworthiness standards: transport category rotorcraft
31	Airworthiness standards: manned free balloons
33	Airworthiness standards: aircraft engines
35	Airworthiness standards: propellers
36	Noise standards
37	Technical standard order (TSO) authorizations
39	Airworthiness directives
43	Maintenance, preventive maintenance, rebuilding, and alteration
45	Identification and registration marking
47	Aircraft registration
49	Recording of aircraft titles and security documents
61	Certification: pilots and flight instructors
65	Certification: airmen other than flight crewmembers
67	Medical standards and certification
91	General operating and flight rules
93	Special air traffic rules and airport traffic patterns
121	Certification and operation: airlines and commercial operators of large aircraft
123	Certification and operation: air travel clubs using large airplanes
135	Air taxi operators and commercial operators of small aircraft
141	Flight schools

Some FARs are sold on a subscription basis by the **Superintendent of Documents, U.S. Government Printing Office, Washington, DC 20402.**

Advisory Circulars (ACs) are information issued by the FAA to inform the aviation public in a systematic way of non-regulatory information of interest (that's a quote). Some ACs are free, others cost money. Three times a year the FAA issues a booklet called *Advisory Circular Checklist,* which lists the status of both FARs and Advisory Circulars. In addition, the FAA issues AC 00-44 *Status of FARs* at regular intervals. This and the *AC Checklist* may be obtained free from **US DOT, Publications Section M443.1, Washington, DC 20590.** ACs are issued about various areas of interest and are numbered according to the pertinent FARs as follows:

Subject Number	*Subject Matter*
00	General
10	Procedural
20	Aircraft
40	Maintenance
60	Airmen
70	Airspace
90	Air traffic control and general operations
120	Air carrier and commercial operators, and helicopters
140	Schools and other certified agencies
150	Airports
170	Air navigation facilities
180	Administrative
210	Flight information

If you want to be placed on an AC mailing list, make the request through **US DOT, Distribution Requirements Section M 482.3, Washington, DC 20590.** Separate mailing lists are maintained for the various areas of interest, so make sure you identify what you want. The best way to do this is to start with a current copy of the *Advisory Circular Checklist,* mentioned above. You can often pick these up at Flight Service Stations.

Notice of Proposed Rule-Making (NPRM)

The FAA uses a process called "Notice of Proposed Rule-Making" to announce possible changes in regulations. Generally, interested parties have 60 days in which to comment on the proposal. NPRM's are published in the Federal Register. You can arrange to receive these notices by contacting **The Department of Transportation (Distribution Requirements Section, TAD 482.3, Washington, DC 20590)** and asking to be put on the mailing list. To receive a specific NPRM, contact the **FAA, Office of Information Services, Attention: Public Information Center, AIS-230, 800 Independence Avenue SW, Washington, DC 20591.** Be sure to give your address, including zip code, and your affiliation with any organization, if applicable. When requesting a specific NPRM, cite the docket number, notice number or other reference to identify it properly.

Schools

There are plenty of learning sources in the aviation field, which is a good thing because we need all the pilots and mechanics we can get. These institutions include:

- Accredited, degree-granting places of higher learning offering specific aviation degrees
- Aviation courses at regular colleges
- Military academies that are branches of the armed services
- Military ROTC courses offered through colleges

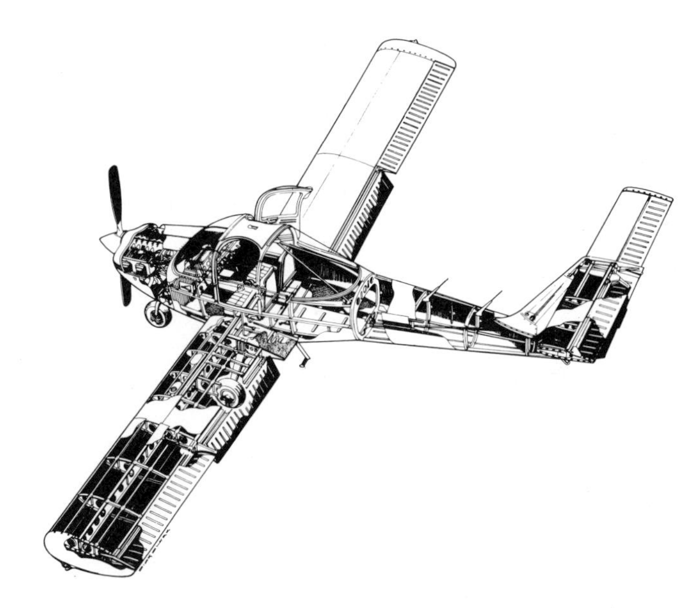

You can learn all about it in a Piper Tomahawk. (Piper drawing)

- National commercial schools offering flight and mechanic training
- Local commercial flying schools offering flight training
- Local non-commercial flying clubs offering flight training
- Specialized cram courses that take you through the material necessary to pass one of the FAA exams
- Aerospace education courses at high schools
- Youth education programs offered by the Boy Scouts of America, the Civil Air Patrol, the 4-H, The Ninety-Nines (the association of women pilots), the Experimental Aircraft Association, and others.

A good source of information on education is the **American Society for Aerospace Education, Suite 1303, 1750 Pennsylvania Avenue NW, Washington, DC 20006, telephone (202) 347-5187.** They publish *Aviation/Space Magazine (The Journal of Aerospace Education),* which includes a biennial directory issue.

Aviation Education Resource Centers are maintained at about 100 universities and colleges throughout the United States. These provide two basic services—they serve as libraries of aviation and space information representing the entire aviation and space community, and they have experienced and knowledgeable directors to provide advice and counseling. General aviation, airlines, manufacturers, and government agencies have all provided materials for these centers.

Among the various institutions offering youth education programs are the Boy Scouts of America, with its Air Explorers (contact **National Director, Aviation Exploring, Boy Scouts of America, North Brunswick, NJ 08902, telephone (201) 249-6000),** and the Civil Air Patrol (CAP), which has such matters as part of its charter. For more information about CAP programs, contact your local CAP Wing, or the CAP Region for the state where you live, via the CAP Director of Aerospace Education as follows:

If you live in:

Connecticut, Massachusetts, Maine, New Hampshire, New Jersey, New York, Pennsylvania, Rhode Island, Vermont

Contact:

USAF-CAP NE Liaison Region
Box 9
Garden City, NY 11530

If you live in:

Delaware, DC, Maryland, North Carolina, South Carolina, Virginia, West Virginia

Contact:

USAF-CAP Middle East Liaison Region
Andrews AFB
MD 20331

If you live in:

Illinois, Indiana, Kentucky, Michigan, Ohio, Wisconsin

Contact:

USAF-CAP Great Lakes Liaison Region
Attention: MCLGLR
Wright-Patterson AFB
OH 45433

If you live in:

Iowa, Kansas, Minnesota, Missouri, Nebraska, North Dakota, South Dakota

Contact:

USAF-CAP North Central Liaison Region
Building 751
Minneapolis-St. Paul International Airport
Minneapolis, MN 55450

If you live in:

Arizona, Arkansas, Louisiana, New Mexico, Oklahoma, Texas

Contact:

USAF-CAP Southwest Liaison Region
USNAS Building 1239
Dallas, TX 75211

If you live in:

Colorado, Idaho, Montana, Utah, Wyoming

Contact:

USAF-CAP Rocky Mountain Liaison Region
Lowry AFB, CO 80230

If you live in:

Alabama, Florida, Georgia, Mississippi, Puerto Rico, Tennesse

Contact:

USAF-CAP Southeast Liaison Region
Box 3117
Dobbins AFB
GA 30060

If you live in:

Alaska, California, Hawaii, Nevada, Oregon, Washington

Contact:

USAF-CAP Pacific Liaison Region
Mather AFB
CA 95655

The 4-H maintains an aerospace education program, with over 20,000 members enrolled. For more information, contact the **4-H Aerospace Program Director, Purdue University, West Lafayette, IN 49706.**

The Ninety-Nines, the international organization of women pilots, runs the Air Age Education Program, which provides flight experience, lectures, and courses for both adult and youth groups. Contact the **Chairman of the International Air Age Committee, Ninety Nines, Box 45021, Dallas, TX 75245.**

Project Schoolflight is a program of the Experimental Aircraft Association Air Museum Foundation aimed at promoting the building of aircraft within schools, scouting organizations, and other youth groups. About 200 aircraft are being built, with over 80 completed. Contact **Project Schoolflight, EAA Air Museum Foundation, Box 229, Hales Corners, WI 53130, telephone (414) 425-4860.**

State Aviation Agencies

Most states have an aeronautical agency of some kind. Some of these are good, actually promoting aviation, improving navaids, building airports, and such. See page 167 for those states that publish aeronautical charts and/or airport directories. Other agencies are strictly just another ripoff, providing an opportunity to gouge the aircraft owner for a meaningless "aircraft registration fee." I leave you to be the judge as to the efficacy of your own state agency, if you have one.

Alabama Department of Aeronautics
Room 627
State Highway Building
11 South Union Street
Montgomery, AL 36130
(205) 832-6290

Alaska Department of Transportation and Public Facilities
Pouch 6900
Anchorage, AK 99502
(907) 266-1470

Arizona Department of Transportation
Aeronautics Division
205 South 17th Avenue
Phoenix, AZ 85007
(602) 261-7778

Arkansas Division of Aeronautics
Adams Field
Old Terminal Building
Little Rock, AR 72202
(501) 376-6781

California Transportation Committee
Aeronautics Subcommittee
1120 N Street
Sacramento, CA 95814
(916) 445-1690

Colorado does not have an aviation agency)

Connecticut Department of Transportation
Bureau of Aeronautics
Drawer A
Wethersfield, CT 06109
(203) 566-4598

Delaware Transportation Authority
Aeronautics Section
Box 778
Dover, DE 19901
(302) 678-4593

Florida Department of Transportation
Aviation Bureau
605 Suwanee Street
Talahassee, FL 32304
(904) 488-8444

Georgia Department of Transportation
Bureau of Aeronautics
5025 New Peachtree Road NE
Chamblee, GA 30341
(404) 393-7353

Hawaii State Department of Transportation
Honolulu International Airport
Honolulu, HI 97819
(808) 847-9432

Idaho Division of Aeronautics and Public Transportation
3483 Rickenbacker Street
Boise, ID 83705
(208) 384-3184

Illinois Division of Aeronautics
Capital Airport
Springfield, IL 62706
(217) 782-2880

Indiana Aeronautics Commission
Suite 801
100 North Senate Avenue
Indianapolis, IN 46204
(317) 633-6545

Iowa Department of Transportation
Aeronautics Division
State House
Des Moines, IA 50319
(515) 281-4280

Kansas Department of Transportation
State Office Building
Topeka, KS 66612
(913) 296-3566

Kentucky Department of Transportation
Division of Aeronautics
419 Ann Street
Frankfort, KY 40601
(502) 564-4480

Louisiana Department of Transportation
Office of Aviation
Box 44245
Capitol Station
Baton Rouge, LA 70804
(504) 342-7504

Maine Department of Transportation
Bureau of Aeronautics
Transportation Building
Augusta, ME 04330
(207) 289-3185

Maryland Department of Transportation
State Aviation Administration
Box 8766
Baltimore-Washington International Airport
MD 21240
(301) 787-7060

Massachusetts Aeronautics Commission
Boston-Logan Airport
East Boston, MA 02128
(617) 727-5350

Michigan Aeronautics Commission
Capital City Airport
Lansing, MI 48906
(517) 373-1834

Minnesota Department of Transportation
Aeronautics Division
Box 417
Transportation Building
St. Paul, MN 55155
(612) 296-8202

Mississippi Aeronautics Commission
Box 5
Jackson, MS 39205
(601) 354-7494

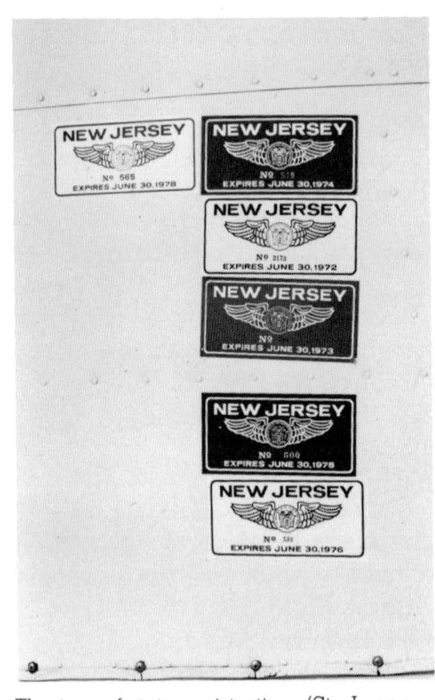

The joys of state registration. (Stu Leventhal photo)

Missouri Department of Transportation
Aviation Section
Box 1250
Jefferson City, MO 65101
(314) 751-4922

Montana Board of Aeronautics
Box 5178
Helena, MT 59601
(406) 449-2506

Nebraska Department of Aeronautics
Box 82088
Lincoln, NE 68501
(402) 471-2371

(Nevada does not have an aviation agency)

New Hampshire Aeronautics Commission
Municipal Airport
Concord, NH 03301
(603) 271-2551

New Jersey Department of Transportation
Division of Aeronautics
1035 Parkway Avenue
Trenton, NJ 08625
(609) 292-3020

New Mexico Department of Transportation
Aviation Division
Box 579
Santa Fe, NM 87503
(505) 827-5511

New York State Department of Transportation
1220 Washington Avenue
Albany, NY 12232
(518) 457-2820

North Carolina Department of Transportation
Division of Aviation
Box 25201
Raleigh, NC 27611
(919) 733-2491

North Dakota Aeronautics Commission
Box U, Municipal Airport
Bismark, ND 58501
(701) 224-2748

Ohio Department of Transportation
Division of Aviation
2829 West Granville Road

Worthington, OH 43085
(614) 889-2533

Oklahoma Aeronautics Commission
424 United Founders Tower Building
Oklahoma City, OK 73112
(405) 521-2377

Oregon State Department of Transportation
Aeronautics Divison
3040 25th Street SE
Salem, OR 97310
(503) 378-4880

Pennsylvania Department of Transportation
Bureau of Aviation
Capital City Airport
New Cumberland, PA 17070
(717) 787-8754

Puerto Rico Ports Authority
Aviation Department
Box 2829
San Juan, PR 00936
(809) 723-2260

Rhode Island Department of Transportation
T. F. Green State Airport
Warwick, RI 02886
(401) 737-4000

South Carolina Aeronautics Commission
Box 1769
Columbia, SC 29202
(803) 758-2766

South Dakota Department of Transportation
Division of Aeronautics
Pierre, SD 57501
(605) 773-3574

Tennessee Department of Transportation
Bureau of Aeronautics
Box 17326
Nashville, TN 37217
(615) 741-3208

Texas Aeronautics Commission
Box 12607
Capitol Station
Austin, TX 78711
(512) 475-4768

Utah Department of Transportation
Division of Aeronautical Operations
135 N 2400 W
Salt Lake City, UT 84116
(801) 328-2066

Vermont Agency of Transportation
133 State Street
Montpelier, VT 05602
(802) 828-2828

Virginia State Corporation Commission
Division of Aeronautics
Box 7716
Richmond, VA 23231
(804) 786-3685

Washington State Department of Transportation
Division of Aeronautics
Boeing Field
Seattle, WA 98108
(206) 764-4131

West Virginia State Aeronautics Commission
Kanawha Airport
Charleston, WV 25311
(304) 348-2689

Wisconsin Department of Transportation
Bureau of Aeronautics
Box 7914
Madison, WI 53707
(608) 266-3351

Wyoming Aeronautics Commission
Cheyenne, WY 82002
(307) 777-7481

Beechcraft Duchess. (Beech photo)

■ PHONETIC ALPHABET AND MORSE CODE

Alfa	•–
Bravo	–•••
Charlie	–•–•
Delta	–•••
Echo	•
Foxtrot	••–•
Golf	––•
Hotel	••••
India	••
Juliett	•–––
Kilo	–•–
Lima	•–••
Mike	––
November	–•
Oscar	–––
Papa	•––•
Quebec	––•–
Romeo	•–•
Sierra	•••
Tango	–
Uniform	••–
Victor	•••–
Whiskey	•––
Xray	–••–
Yankee	–•––
Zulu	––••

Piper Aerostar 600A. (Piper photo)

Key to Airspeed Terminology

Airspeeds are referred to in mysterious ways by pilots, aircraft designers, the FAA, and others. Here is the key:

Symbol	Meaning
IAS	Indicated airspeed—the speed of an airplane shown on the airspeed indicator.
CAS	Calibrated airspeed—the indicated airspeed, corrected for position and instrument error.
TAS	True airspeed—the airspeed of an airplane in undisturbed air, which is the CAS corrected for altitude, temperature, and compressibility error.
GS	Ground speed—the speed of an airplane relative to the ground.
M	Mach number—the ratio of the true airspeed to the speed of sound. M 1.0 is the speed of sound, M 0.8 is 80 percent of the speed of sound, etc.
V_1	Decision speed in the event of an engine failure on takeoff—in a multi-engine aircraft, the takeoff may be either rejected or continued at this speed. Below this speed it is rejected. Above this speed it is continued.
V_2	Takeoff safety speed—the lowest speed at which the airplane complies with handling criteria associated with the climb after takeoff following an engine failure.
V_A	Maneuvering speed—the maximum speed at which application of full available aerodynamic control will not overstress the airplane.
V_{DF} M_{DF}	Maximum demonstrated diving speed—the highest speed demonstrated during certification (V expressed in knots, M expressed as a Mach number).
V_F	Maximum flap-operating speed—the maximum speed at which flaps may be extended or retracted.
V_{FE}	Maximum flap-extended speed—the highest speed permissible with wing flaps in a prescribed extended position.
V_{LE}	Maximum landing-gear-extended speed—the maximum speed at which the airplane can be safely flown with the gear down.
V_{LO}	Maximum landing-gear-operating speed—the maximum speed at which the landing gear can be safely extended or retracted.
V_{MCA}	Minimum control-speed in the air in take-off configuration—also the minimum speed at which it is possible to suffer an engine failure in a multi-engine aircraft and still maintain control of the aircraft within defined limits.
V_{MCG}	Minimum control speed on the ground—the minimum speed at which it is possible to suffer an engine failure on takeoff in a multi-engine aircraft and maintain control of the airplane within defined limits.

Symbol	Meaning
V_{NE} M_{NE}	Never exceed speed—the speed that may not be exceeded at any time (V = knots, M = Mach number).
V_{NO} M_{NO}	Maximum structural cruising speed—the speed that should not be exceeded except in smooth air and then only with caution (V = knots, M = Mach number).
V_R	Rotation speed—the speed at which the pilot starts to rotate the aircraft for takeoff.
V_S	The stall speed, or the minimum steady-flight speed at which the aircraft is controllable.
V_{SO}	The stall speed, or the minimum steady flight speed at which the airplane is controllable in the landing configuration.
V_{SSE}	The recommended safe one-engine-inoperative speed—this provides a margin of safety above V_{MCA}.
V_X	Best angle-of-climb speed—the airspeed that delivers the greatest gain of altitude in the shortest possible horizontal distance.
V_Y	Best rate-of-climb speed—the airspeed that delivers the greatest gain of altitude in the shortest possible time.

Abbreviations

AAF	Army Air Field
AAL	Above Airport Level
AAS	Airport Advisory Service
ABM	Abeam
AC	Advisory Circular
ACC	Area Control Center
AD	Airworthiness Directive
ADCUS	Advise Customs
ADF	Automatic Direction Finder
ADIZ	Air Defense Identification Zone
A&E	Airframe and Engine Mechanic
AFB	Air Force Base
AGL	Above Ground Level
AH	Artificial Horizon
AIM	Airman's Information Manual
AIRMET	Light Aircraft Weather Warning
ALS	Approach Light System
ALT	Altitude
ALT	Alternate
AMSL	Above Mean Sea Level
A&P	Airframe and Powerplant Mechanic
AP	Autopilot
ARINC	Aeronautical Radio Inc.
ARP	Airport Reference Point
ARSR	Air Route Surveillance Radar
ARTCC	Air Route Traffic Control Center
ASI	Airspeed Indicator
ASL	Above Sea Level
ASR	Airport Surveillance Radar
ATA	Actual Time of Arrival
ATC	Air Traffic Control
ATCC	Air Traffic Control Center
ATCRBS	ATC Radar Beacon System
ATD	Actual Time of Departure
ATIS	Automatic Terminal Information Service
BC	Back Course
BCN	Beacon
BCST	Broadcast
BHP	Brake Horsepower
BKN	Broken (clouds)
BM	Back Marker
CAB	Civil Aeronautics Board
CADIZ	Canadian Air Defense Identification Zone
CAS	Calibrated Air Speed
CAT	Carburetor Air Temperature Gauge
CAT	Category
CAT	Clear Air Turbulence
CGAS	Coast Guard Air Station
CHT	Cylinder Head Temperature Gauge
CMOH	Chrome Major Overhaul
COM	Communications Radio
CTOH	Chrome Top Overhaul
CVFR	Controlled VFR
DCT	Direct
DEWIZ	Distant Early Warning Identification Zone
DF	Direction Finding
DG	Directional Gyro
DH	Decision Height
DME	Distance Measuring Equipment
DOT	Department of Transportation
DR	Dead Reckoning
DVOR	Doppler VOR
EDT	Eastern Daylight Time
EFAS	Enroute Flight Advisory Service
EGT	Exhaust Gas Temperature Gauge
ELT	Emergency Locator Transmitter
EMDO	Engineering & Manufacturing District Office
EST	Eastern Standard Time
ETA	Estimated Time of Arrival

ETD	Estimated Time of Departure
ETE	Estimated Time Enroute
FAA	Federal Aviation Administration
FAF	Final Approach Fix
FD	Flight Director
FIR	Flight Information Region
FL	Flight Level
FM	Fan Marker
FPM	Feet Per Minute
FREMAN	Factor Remanufactured Engine
FREQ	Frequency
FSS	Flight Service Station
FT	Feet
FTO	Ferry Time Only
FWFWD	Firewall Forward
GADO	General Aviation District Office
GCA	Ground Controlled Approach
GMT	Greenwich Mean Time
GPH	Gallons Per Hour
GS	Glide Slope
GS	Ground Speed
HAA	Height Above Airport
HDG	Heading
HF	High Frequency (3-30 MHz)
HIRL	High Frequency Runway Lights
HSI	Horizontal Situation Indicator
Hz	Hertz (cycles per second)
IAF	Initial Approach Fix
IAS	Indicated Airspeed
IATA	International Air Transport Association
ICAO	International Civil Aviation Organization
IDENT	Identification
IFR	Instrument Flight Rules
IFSS	International Flight Service Station
ILS	Instrument Landing System
IM	Inner Marker
IMC	Instrument Meteorological Conditions
INS	Inertial Navigation System
IRAN	Inspected and Repaired As Needed
ISA	International Standard Atmosphere
IVSI	Instantaneous Vertical Speed Indicator
JATO	Jet-Assisted Takeoff
KHz	KiloHertz
KM	Kilometer
KMH	Kilometers Per Hour
KT	Knot (nautical mile per hour)
LAT	Latitude
LDA	Localizer Directional Aid
LDIN	Lead-In Lighting System
LF	Low Frequency (30-300 KHz)
LFR	Low Frequency Range
LIM	Locator Inner Marker
LMM	Locator Middle Marker
LOC	Localizer
LOC	Locator
LOM	Locator Outer Marker
LONG	Longitude
MAA	Maximum Authorized Altitude
MAG	Magnetic
MAP	Missed Approach Point
MAX	Maximum
MB	Marker Beacon
MB	Millibars
MCA	Minimum Crossing Altitude
MCAS	Marine Corps Air Station
MDA	Minimum Descent Altitude
MDH	Minimum Decision Height
MEA	Minimum Enroute Altitude
MF	Medium Frequency (300-3000 KHz)
MHz	MegaHertz
MIN	Minimum
MIN	Minute
MKR	Marker Beacon
MLS	Microwave Landing System
MM	Middle Marker
MOA	Military Operation Area
MOCA	Minimum Obstruction Clearance Altitude
MOT	Ministry of Transport (Canada)
MPH	Miles Per Hour
MRA	Minimum Reception Altitude
MSA	Minimum Safe Altitude
MSL	Mean Sea Level
MUN	Municipal
MVFR	Marginal VFR
NAS	Naval Air Station
NAV	Navigational Aid/Receiver
NAVAID	Navigational Aid
NDB	Non-Directional Beacon
NDH	No Damage History
NM	Nautical Miles
NOAA	National Oceanic & Atmospheric Administration
NOTAM	Notice To Airmen
NWS	National Weather Service
OAT	Outside Air Temperature
OBS	Omni Bearing Selector
OM	Outer Marker
OVCST	Overcast (clouds)
PAR	Precision Approach Radar
PATWAS	Pilot's Automatic Telephone-Answering Service
PCPN	Precipitation
PIREP	Pilot Report (weather)
PPO	Prior Permission Only
PWI	Proximity Warning Indicator
QDM	Magnetic Bearing to Facility
QDR	Magnetic Bearing from Facility
QFE	Height Above Ground Based on Local Station Pressure

QNE	Altimeter Setting 29.92 Hg or 1013.2 MB
QNH	Altitude Above Mean Sea Level Based on Local Station Pressure
RAPCON	Radar Approach Control
RB	Rotating Beacon
RC	Rate of Climb
RDR	Radar
RE	Right Engine
REIL	Runway End Identification Lights
REMAN	Remanufactured
RMI	Radio Magnetic Indicator
RNAV	Area Navigation
RNG	LF Radio Range
RPM	Revolutions Per Minute
RVR	Runway Visual Range
RVV	Runway Visibility Value
RWY	Runway
SCMOH	Since Chrome Major Overhaul
SCTOH	Since Chrome Top Overhaul
SCTD	Scattered (clouds)
SDF	Simplified Directional Facility
SEC	Seconds
SEC	Sectional Aeronautical Chart
SFL	Sequenced Flashing Lights
SFREMAN	Since Factory Remanufacture
SID	Standard Instrument Departure
SIGMET	Significant Meteorological Information
SM	Statute Miles
SMOH	Since Major Overhaul
SPOH	Since Propeller Overhaul
SR-SS	Sunrise to Sunset
STAR	Standard Terminal Arrival Route
STC	Supplemental Type Certificate
STD	Standard
STOH	Since Top Overhaul
STOL	Short Takeoff and Landing
SWB	Scheduled Weather Broadcast
TACAN	Tactical Air Navigation
TAS	True Airspeed
TASI	True Airspeed Indicator
TB	Turn and Bank
TBO	Time Between Overhauls
TC	Turn Coordinator
TC	Type Certificate
TCA	Terminal Control Area
TRSA	Terminal Radar Service Area
TSO	Technical Standard Order
TT	Total Time
TTAE	Total Time Airframe and Engines
TTAF	Total Time Airframe
TTSN	Total Time Since New
TVOR	Terminal VOR
TWEB	Transcribed Weather Broadcast
TWR	Tower
TXP	Transponder
UFN	Until Further Notice
UHF	Ultra High Frequency (300-3000 MHz)
UNICOM	Aeronautical Advisory Frequency
USAF	U.S. Air Force
USN	U.S. Navy
VAR	Variation
VAR	Visual Aural Range
VASI	Visual Approach Slope Indicator
VDF	VHF Direction Finding
VDP	Visual Descent Point
VFR	Visual Flight Rules
VHF	Very High Frequency (30-300 MHz)
VLF	Very Low Frequency (3-30 KHz)
VMC	Visual Meteorological Conditions
VNAV	RNAV with Vertical Guidance
VOR	VHF Omnidirectional Radio Range
VOR/LOC	VOR and Localizer
VORTAC	VOR and TACAN Co-located
VOT	VOR Test Signal
VSI	Vertical Speed Indicator
VTOL	Vertical Takeoff and Landing
VV	Vertical Visibility
WAC	World Aeronautical Chart
WILCO	Will Comply
WP	Waypoint (area navigation)
XMTR	Transmitter
XPDR	Transponder
XTAL	Crystal
Z	Greenwich Mean Time (Zulu)

5 F4Us, 1944- (US Navy photo)

International Civil Aircraft Markings

Marking	Country
A-2	Botswana
A6	United Emirates
A7	Qatar
A9	Bahrain
A40	Oman
AN	Nicaragua
AP	Pakistan
B	China (People's Republic)
B	Taiwan
C-2	Nauru
C5	Gambia
C6	Bahamas
C9	Mozambique
C	Canada
CC	Chile
CCCP	USSR
CN	Morocco
CP	Bolivia
CR-C	Cape Verde Islands
CS	Portugal
CU	Cuba
CX	Uruguay
D2	Angola
D6	Comoro Islands
D	Germany (Federal Republic)
DM	East Germany
DQ	Fiji
EC	Spain
EI	Ireland
EL	Liberia
EP	Iran
ET	Ethiopia
F	France
F-O	French Protectorates
G	United Kingdom
H4	Solomon Islands/New Hebrides
HA	Hungary
HB	Switzerland
HB	Liechtenstein
HC	Ecuador
HH	Haiti
HI	Dominican Republic
HK	Colombia
HL	South Korea
HMAY	Mongolia
HP	Panama
HR	Honduras
HS	Thailand
HZ	Saudi Arabia
I	Italy
J5	Guinea Bissau
JA	Japan
JY	Jordan
LN	Norway
LV	Argentina
LX	Luxembourg
LZ	Bulgaria
N	USA
OB	Peru
OD	Lebanon
OE	Austria
OH	Finland
OK	Czechoslovakia
OO	Belgium
OY	Denmark
P2	Papua New Guinea
P	North Korea
PDRL	Laos
PH	Netherlands
PJ	Netherlands Antilles
PK	Indonesia
PK	West Irian
PP	Brazil
PT	Brazil
PZ	Surinam
RP	Philippines
S-2	Bangladesh
S7	Seychelles
S9	Sao Tome
SE	Sweden
SP	Poland
ST	Sudan
SU	Egypt
SX	Greece
TC	Turkey
TF	Iceland
TG	Guatemala
TI	Costa Rica
TJ	Cameroon
TL	Central African Republic
TN	Congo
TR	Gabon
TS	Tunisia
TT	Chad
TU	Ivory Coast
TY	Benin
TZ	Mali
VH	Australia
VN	Vietnam
VP-F	Falkland Islands
VP-H	Belize
VP-L	Leeward Islands
VP-P	Gilbert Islands
VP-V	St. Vincent
VP-W	Zimbabwe Rhodesia
VP-Y	Zimbabwe Rhodesia
VQ-G	Grenada
VQ-L	St. Lucia
VR-B	Bermuda
VR-C	Cayman Islands
VR-G	Gibraltar
VR-H	Hong Kong
VR-U	Brunei
VT	India
XA	Mexico
XB	Mexico
XC	Mexico
XT	Upper Volta
XY	Burma
XZ	Burma
YA	Afghanistan
YI	Iraq
YK	Syria
YR	Romania
YS	El Salvador
YU	Yugoslavia
YV	Venezuela
ZA	Albania
ZK	New Zealand
ZP	Paraguay

Venezuelan and U.S. King Airs. (Beech photo)

ZS	South Africa
3A	Monaco
3B	Mauritius
3D	Swaziland
3X	Guinea
4R	Sri Lanka
4W	Yemen
4X	Israel
5A	Libya
5B	Cyprus
5C	Equatorial Guinea
5H	Tanzania
5N	Nigeria
5R	Malagasy Republic
5T	Mauritania
5U	Niger
5V	Togo
5W	Western Samoa
5X	Uganda
5Y	Kenya
5Y	Kenya
6O	Somalia
6V	Senegal
6Y	Jamaica
7O	Republic of Yemen
7P	Lesotho
7Q	Malawi
7T	Algeria
8P	Barbados
8Q	Maldive Republic
8R	Guyana
9G	Ghana
9H	Malta
9J	Zambia
9K	Kuwait
9L	Sierra Leone
9M	Malaysia
9N	Nepal
9Q	Zaire
9U	Burundi
9V	Singapore
9XR	Rwanda
9Y	Trinidad and Tobago

Algerian Beech Queen Airs 7T-VIB and 7T-VIC. (Beech photo)

International Civil Aircraft Markings

France's Falcon 50, F-WAMD. (Falcon Jet photo)

Afghanistan	YA
Albania	ZA
Algeria	7T
Angola	D2
Argentina	LV
Australia	VH
Austria	OE
Bahamas	C6
Bahrain	A9
Bangladesh	S-2
Barbados	8P
Belgium	OO
Belize	VP-H
Benin	TY
Bermuda	VR-B
Bolivia	CP
Botswana	A-2
Brazil	PP
Brazil	PT
Brunei	VR-U
Bulgaria	LZ
Burma	XY
Burma	XZ
Burundi	9U
Cameroon	TJ
Canada	C
Cape Verde Islands	CR-C
Cayman Islands	VR-C
Central African Republic	TL
Chad	TT
Chile	CC
China (People's Republic)	B
Colombia	HK
Comoro Islands	D6
Congo	TN
Costa Rica	TI
Cuba	CU
Cyprus	5B
Czechoslovakia	OK
Denmark	OY
Dominican Republic	HI
East Germany	DM
Ecuador	HC
Egypt	SU
El Salvador	YS
Equatorial Guinea	5C
Ethiopia	ET
Falkland Islands	VP-F
Fiji	DQ
Finland	OH
France	F
French Protectorates	F-O
Gabon	TR
Gambia	C5
Germany (Federal Republic)	D
Ghana	9G
Gibraltar	VR-G
Gilbert Islands	VP-P
Greece	SX
Grenada	VQ-G
Guatemala	TG
Guinea	3X
Guinea Bissau	J5
Guyana	8R
Haiti	HH
Honduras	HR
Hong Kong	VR-H
Hungary	HA
Iceland	TF
India	VT
Indonesia	PK
Iran	EP
Iraq	YI
Ireland	EI
Israel	4X
Italy	I
Ivory Coast	TU
Jamaica	6Y
Japan	JA
Jordan	JY
Kenya	5Y
Kenya	5Y
Korea (North)	P
Korea (South)	HL
Kuwait	9K
Laos	PDRL
Lebanon	OD
Leeward Islands	VP-L
Lesotho	7P
Liberia	EL
Libya	5A
Liechtenstein	HB
Luxembourg	LX
Malagasy Republic	5R
Malawi	7Q
Malaysia	9M
Maldive Republic	8Q
Mali	TZ
Malta	9H
Mauritania	5T
Mauritius	3B
Mexico	XA
Mexico	XB
Mexico	XC
Monaco	3A
Mongolia	HMAY
Morocco	CN
Mozambique	C9
Nauru	C-2
Nepal	9N
Netherlands	PH
Netherlands Antilles	PJ
New Zealand	ZK
Nicaragua	AN
Niger	5U
Nigeria	5N
Norway	LN
Oman	A40
Pakistan	AP
Panama	HP
Papua New Guinea	P2
Paraguay	ZP
Peru	OB
Philippines	RP
Poland	SP
Portugal	CS
Qatar	A7
Republic of Yemen	7O
Romania	YR

Rwanda	9X
Sao Tome	S9
Saudi Arabia	HZ
Senegal	6V
Seychelles	S7
Sierra Leone	9L
Singapore	9V
Solomon Islands/New Hebrides	H4
Somalia	6O
South Africa	ZS
Spain	EC
Sri Lanka	4R
St. Lucia	VQ
St. Vincent	VP-V
Sao Tome Island	S9
Saudi Arabia	HZ
Senegal	6V
Seychelles	S7
Sierra Leone	9L
Singapore	9V
Solomon Islands	H4
Somalia	6O
South Africa	ZS
Soviet Union	CCCP
Spain	EC
Sri Lanka	4R
Sudan	ST
Surinam	PZ
Swaziland	3D
Sweden	SE
Switzerland	HB
Syria	YK
Taiwan	B
Tanzania	5H
Thailand	HS
Togo	5V
Transkei	ZS
Trinidad and Tobago	9Y
Tunisia	TS
Turkey	TC
Uganda	5X
United Arab Emirates	A6
United States of America	N
Upper Volta	XT
Uruguay	CX
Venezuela	YV
Vietnam	VN
Virgin Islands (British)	VP-L
Western Samoa	5W
Yemen Arab Republic	4W
Yemen PDR	7O
Yugoslavia	YU
Zaire	9Q
Zambia	9J
Zimbabwe (Rhodesia)	VP-W, VP-Y

Form Approved: OMB No. 04-R0072

DEPARTMENT OF TRANSPORTATION
FEDERAL AVIATION ADMINISTRATION

FLIGHT PLAN

CIVIL AIRCRAFT PILOTS. FAR Part 91 requires you file an IFR flight plan to operate under instrument flight rules in controlled airspace. Failure to file could result in a civil penalty not to exceed $1,000 for each violation (Section 901 of the Federal Aviation Act of 1958, as amended). Filing of a VFR flight plan is recommended as a good operating practice. See also Part 99 for requirements concerning DVFR flight plans.

1. TYPE: VFR / IFR / DVFR
2. AIRCRAFT IDENTIFICATION
3. AIRCRAFT TYPE/ SPECIAL EQUIPMENT
4. TRUE AIRSPEED — KTS
5. DEPARTURE POINT
6. DEPARTURE TIME — PROPOSED (Z) / ACTUAL (Z)
7. CRUISING ALTITUDE
8. ROUTE OF FLIGHT
9. DESTINATION (Name of airport and city)
10. EST. TIME ENROUTE — HOURS / MINUTES
11. REMARKS
12. FUEL ON BOARD — HOURS / MINUTES
13. ALTERNATE AIRPORT(S)
14. PILOT'S NAME, ADDRESS & TELEPHONE NUMBER & AIRCRAFT HOME BASE
15. NUMBER ABOARD
16. COLOR OF AIRCRAFT

CLOSE VFR FLIGHT PLAN WITH____________FSS ON ARRIVAL

FAA Form 7233-1 (8-77)

Rockwell Commander 700.(Rockwell photo)

INDEX

Page numbers in *italics* indicate illustrations. Page numbers in **bold** face type indicate addresses.

Hughes 269A. (Hughes photo)